PRAISE FOR

Mirage: Napoleon's Scientists and the Unveiling of Egypt

"Burleigh's description of a young army overdressed for the sweltering heat (in Alpine wool uniforms), afraid and unable to communicate with the increasingly hostile locals, has echoes of the present. Her principal subject, however, is not the military but the 151 'savants' Napoleon took along—geologists, mapmakers, naturalists, artists, even a musicologist. . . . In less than 250 pages Burleigh hurtles through the three grueling years the savants spent in Egypt, peppering her tale with multitudes of facts, digressions, and anecdotes."

—*New York Times Book Review* (Editors' Choice)

"With an easy style and an eye for striking detail, Burleigh concentrates on 151 French scientists, scholars, and students who joined the expedition tempted by hero-worship of Napoleon and the prospect of scientific adventure."　　　　—Associated Press

"Burleigh . . . explains significant details without getting heavily academic. By separating the narrative into sections and sketching individuals—the chemist, the mathematician, the zoologist—she makes the discussion accessible. . . . A fascinating read about an extraordinary time and place in world history."

—*San Francisco Chronicle*

"*People* staff writer Nina Burleigh spotlights the Indiana Jones–esque scientists who joined Napoleon's Egyptian invasion during the late eighteenth century."　　　　　　　　　　　—*People*

"This fascinating book, covering a catastrophic nineteenth-century attempt to force Western democracy onto the Arab world, is a stark reminder that there is nothing new about jihadism or violent opposition to unwelcome Western intervention. The book is both engrossing and enraging."

—Arianna Huffington, Editor in Chief, Huffingtonpost.com

"Burleigh . . . offers an absorbing glimpse of Napoleon's thwarted bid for a grand French empire and its intellectual fruits."

—*Publishers Weekly*

"If you enjoy delving into small crevices of the past looking for little-considered gems of history, then Burleigh's . . . latest is for you. Focusing on Napoleon's expedition to Egypt in 1798–1801, and particularly on the scientists who accompanied the military forces, Burleigh illuminates an unfamiliar moment in the history of science. . . . Burleigh's storytelling ability is mesmerizing; she skillfully fills in the backstory of the region in artfully crafted paragraphs, summing up thousands of years of history without slowing the flow of the narrative."
                                                                    —*Library Journal*

"Plague and war hampered [the French expedition's] activity, and death eventually carried off dozens of their number in the years they were stranded—by Nelson's destruction of Napoleon's fleet—in Egypt. These privations stand out in Burleigh's narrative, bestowing the savants with a scientific heroism that Burleigh wraps in the ambiguities of the attempted conquest of which they were an integral part. Better than a good story well told, Burleigh's is a perceptive appraisal of this fateful interaction between Western science and Middle Eastern culture."
                                                                    —*Booklist*

"Burleigh's account captures the personalities of the scholars involved as they cope with both the culture shock of surviving in a foreign land and the difficulties of adapting to life in the French military. . . . Burleigh's book succeeds . . . bringing to life a bygone era that still has much to teach us."
                                                                    —*Archaeology* magazine

"[Burleigh] tells a lively story. . . . In her closing chapters, [Burleigh] vividly describes the nightmare that had befallen the occupiers. Devastated by plague, the French fitfully retreated in late 1801 as the English plundered the savants' finds. Holed up in a besieged Alexandria, a bedraggled, emaciated Saint-Hilaire deliriously contemplated a unifying theory of life, 'a principle so gigantic it unified all the sciences,' as he dissected an electric fish. Writing with manic energy as bombs exploded outside, Saint-Hilaire mused on 'the imponderable fluids' of light, electricity, and heat as he tried to pinpoint a link to 'all the phenomena of the material world.' It is a stunning image and a fitting metaphor for the overreaching ambition that drove the savants in their quest."
                                                                    —*New York Sun*

About the Author

NINA BURLEIGH is the author of two widely praised books, *The Stranger and the Statesman: James Smithson, John Quincy Adams, and the Making of America's Greatest Museum: The Smithsonian* and *A Very Private Woman: The Life and Unsolved Murder of Presidential Mistress Mary Meyer*. Her articles have appeared in *Time*, the *Washington Post*, *New York* magazine, the *New York Observer*, *Elle*, *Mirabella*, *Redbook*, *Jane*, *Spy*, *Regardie's*, the *Chicago Tribune*, *George*, Salon.com, and many other publications. She has traveled widely in the United States, covering American elections, and extensively in the Middle East, reporting from inside Iraq during the 1990s on assignment for *Time*. She lives in New York with her family.

Also by Nina Burleigh

*A Very Private Woman: The Life and Unsolved Murder
of Presidential Mistress Mary Meyer*

*The Stranger and the Statesman:
James Smithson, John Quincy Adams, and the
Making of America's Greatest Museum:
The Smithsonian*

MIRAGE

*Napoleon's Scientists and
the Unveiling of Egypt*

NINA BURLEIGH

HARPER PERENNIAL

NEW YORK • LONDON • TORONTO • SYDNEY • NEW DELHI • AUCKLAND

HARPER ● PERENNIAL

A hardcover edition of this book was published in 2007 by HarperCollins Publishers.

HarperCollins books may be purchased for educational, business, or sales promotional use. For information please write: Special Markets Department, HarperCollins Publishers, 10 East 53rd Street, New York, NY 10022.

FIRST HARPER PERENNIAL EDITION PUBLISHED 2008.

Designed by Emily Cavett Taff

Library of Congress Cataloging-in-Publication Data is available upon request.

ISBN 978-0-06-059768-9

08 09 10 11 12 ID/RRD 10 9 8 7 6 5 4 3 2 1

ACKNOWLEDGMENTS

I am grateful to Howard Cohn and Claire Wachtel for reading and commenting on many versions of this manuscript; Elisabeth Buzay for translating primary documents and providing organizational support; Bob Brier and Pat Remler for generously sharing knowledge and their collection of books and art; David Smith and the New York Public Library; also Erik Freeland, Lisa Bankoff, Lauretta Charlton, Allan Metcalf, Carolyn Waters, Yves Laissus, Robert Burleigh, Berta Burleigh, Marcia Isaacson, Ambassador Mahmoud Allam, Charles Coulston Gillispie, Alfred and Charlotte Kessler, Howeyda El Sayed, Nelly Korbel, Stephanie Slewka, Gwendolen Cates, Elizabeth Robinson, Julia Novitch, Liz McNeil, the technology staff at *People* magazine, and Renata Mlynska. The memory of my late grandmother, Ghazal Thomas, who walked out of the remains of the Ottoman Empire, inspired me.

CONTENTS

INTRODUCTION

A little more than two hundred years ago, Europeans contemplated the Islamic countries of the Middle East from afar and imagined rare silks and spices, harems, and gold—yellow gold, not the underground sea of black gold that modern Westerners associate with the region. The territory was largely unmapped, its history and people almost as obscure as the dark side of the moon. Only the most reckless or mad European had dared traverse the lands of the Prophet in the years between the Crusades and the end of the eighteenth century. The gory conflict between Christianity and Islam still loomed large in the collective memory. Tales of impalements, beheadings, and other atrocities in Islamic Central Europe over a three-hundred-year period, culminating with the Battle of Vienna in the late seventeenth century, were vivid deterrents to the few intrepid souls—merchants and writers, mostly—who would consider the journey. The inhabitants of the Middle East were assumed to be inhospitable, and the climate was known to be extreme. The voyage from Europe to the Orient, as it was then called, was arduous and long, thirty days at least by sea, often more.

That never stopped people from fantasizing about what lay beyond the water. On the contrary, it only increased the fascination. Egypt, and its extinct ancient culture, just across the Mediterranean Sea, had tantalized Europeans for centuries. They knew colossal relics of the oldest-known human civilization were concentrated along the Nile in crumbling piles between two vast, usurping deserts, amidst a modern population that professed faith in Islam. Europeans attached all sorts of inferences to this place, viewing it variously as the primal seat of natural law, the remains of a golden age of civilization, and a repository of lost magical knowledge. Few ever got close enough to really know.

By the end of the eighteenth century, such questions as what these monuments were and who made them had hung over the Nile valley since biblical times. The Egyptians themselves could not explain them. The hieroglyphic script—carved into every inch of space on colossal walls and columns—was believed even by scholars to be comprised not of phonetic symbols but magical formulas capable of reviving the dead or turning lead into gold. European doctors thought ground-up mummies were medicinal. Ignorance, bad scholarship, and faulty memories settled on the ancient sites, along with layers of sand drifting in from the two deserts.

The French did not invade Egypt in 1798, however, to solve historical mysteries. They sought colonial power and commerce, at the dawn of the modern global economy. When Napoleon led 34,000 soldiers and 16,000 sailors across the Mediterranean to the distant desert country, the young general undertook a bold (many said crazy) thrust in the ongoing competition among European countries for influence in distant parts of the globe.

France and England had been vying for economic control of territory, from India to North America to the South Seas, since the 1600s. The benefits of such control were well understood. A new class of fantastically wealthy men were building mansions in England and on the Continent, living like blood nobility off the bounty of colonial commerce. England had lost its American colonies to the Americans, but it still controlled most of India, regarded by Europeans as the greatest possible Asian possession.

Between 1774 and 1798, the French government had entertained at least a dozen proposals from various diplomats, politicians, and businessmen to invade Egypt. "Egypt," as one French diplomat counseled the ill-fated King Louis XVI, "does not belong to anybody." But the time to claim it had never seemed ripe, until the right man came along.

Napoleon's gambit was brazen and ill-timed. The French had just recovered from their Revolution. Their economy was in tatters, civic life barely restored. The streets of Paris ran with sewage, and the city smelled worse than it had during the Middle Ages. The newly anti-monarchist country that had killed its king had been fighting wars with royalist nations for several years. Diverting soldiers and matériel to Egypt in 1798, while European foes still menaced their borders, was hardly a conventional allocation of resources by the leaders of France. They were for it, though, because they longed to realize their share of the benefits of empire.

Napoleon and the French government hoped that taking Egypt would be the first step toward founding a grand French empire that would encompass generous swathes of Africa and Asia. The French had made colonial incursions into Asia, but since the British repulsed them at Bengal in 1757, French influence in India had waned to almost nothing, and French leaders still salivated for a piece of the Asian pie. Egypt, with its Mediterranean coast and distant Turkish government-by-proxy in the form of the Mameluke dynasties, had been a tempting objective for decades.

When the French arrived, the Mamelukes had ruled Egypt for more than five hundred years. In Arabic, the word *mamluk* means "the possessed." The Mamelukes were a bizarre slave caste that had been Islam's elite fighting force for nearly a thousand years. They were white Eurasian men kidnapped or purchased as children and then sold at markets in Damascus, Istanbul, and Cairo expressly to be trained in equestrian fighting and rigorous Islam, in order to defend their masters.

In the ninth century this warrior slave caste—androcentric, ascetic, and Orthodox Muslim—overwhelmed their masters in Baghdad. From then on, although Mamelukes replenished their ranks with white boys purchased from slavers, they were slaves in name only.

The Egyptian campaign had many eccentric, even unbelievable, aspects, starting with the Mamelukes themselves, and including the fact that the 50,000 French soldiers and sailors mustered for the invasion weren't told *where* they were going until their ships were almost within sight of the target. The corps of 151 Parisian artists and scientists organized to accompany the soldiers was another surreal facet of the adventure. Responding to the young general's call for *savants* to help explore a secret destination, a group of Paris's brightest intellectual lights left the safety of their labs, studios, and classrooms and boarded ships. Astronomers, mathematicians, naturalists, physicists, doctors, chemists, engineers, botanists, and artists—even a poet and a musicologist—locked up their desks, packed their books, said goodbye to friends and family, and undertook what was, literally, for most of them, a voyage into the unknown.

Bringing scientists along gave credence to the ideal of this *mission civilisatrice*. Claiming to bring French-style culture and democracy to Arabs ruled by non-Arab tyrants offered moral cover for the invasion. Napoleon also had a classical precedent for bringing scientists on a military campaign. His spiritual role model, Alexander the Great, had traveled with philosophers when he invaded Persia. Having a human encyclopedia at his side added a certain elegance to the brutal endeavor. Besides accomplishing the Enlightenment goals of categorizing and classifying, Napoleon also expected his *savants* to help administer the conquered territory, mapping the land, finding the water, befriending the leaders, and even negotiating with the foe. Some of the civilians did indeed become integral to the military occupation, in terms of support and administration.

The scientists, however, believed they were along primarily to make discoveries and practice science. They were mostly young: their median age was twenty-five. The older men among them were prominent enough to know each other by their membership in the Institute of France. Few were intimate friends when they set sail, but over the course of the next three years, they would form lifelong bonds (and enmities). Those who returned to France alive would spend the rest of their lives arguing over the meaning of what they had seen and experienced, until death picked them off, one by one. In middle age, the younger men eulogized their elder counterparts. Two even asked to be buried in adjoining graves.

The scholars included the most promising and prominent men in the French sciences. And apart from a handful of Orientalists, most knew next to nothing about Egypt. They arrived in Egypt with different specialties, some of which (painting, musicology, geometry, physics) seemed quite superfluous to the making of a French desert colony. Yet such eminences as the esteemed chemist Claude-Louis Berthollet and the geometer Gaspard Monge were already in their fifties and had clearly both the scientific and administrative skills to be very useful to a conquering army. Monge and Berthollet were old hands at overseeing the organization and transport of military "collections" (looted art, mainly from Italy) for Napoleon, and at producing gunpowder for the revolutionary army. Others were young teachers and researchers, recruited straight out of classrooms, moved to join equally by reverence for Napoleon and hopes of career advancement. The thirty-six students who signed on were following adolescent dreams of travel and adventure. Some developed passions in Egypt that influenced the rest of their lives; some never made it back to France.

Those *savants* who survived produced an exhaustive encyclopedia of Egypt, twenty-three outsize volumes, delicately printed with engravings of the buildings, rocks, people, plants, and thousands of the beasts, birds, bugs, and fish that dwelled in Egypt circa 1800. *La Description de l'Égypte* (The Description of Egypt) comprised the ultimate work of the scientists. Their encyclopedia is a record of the impressions of European civilian participants in the first large-scale interaction between Europeans and Muslims in the modern era. Thirty-one of them died in Egypt, or shortly thereafter; many of the rest returned home morally, emotionally, or physically altered for life. The experience was so affecting that until the last of them died in the 1860s, they referred to one another as "the Egyptians" and met annually to reminisce about the sand-drifted statues, temples, and tombs they had measured, sketched, and sought to understand.

When they first got to Egypt, the scientists tried to approach the land, people, animals, and relics not as tourists or literary travelers, or even colonizers, but from within their fields of scholarship. They categorized, measured, and collected, kept journals, and wrote reports that they read to one another at formal meetings in a recently vacated harem room. Set down and then abandoned in what was mostly a vast, uncharted desert, atheists surrounded by devout practitioners of a barely understood religion, encountering ruins that mutely testified to another equally unknown culture, they did not trust to sense perception alone. They looked at and tried to explain what they saw by mathematically defining and classifying it.

In that effort, they founded a new science that would come to be called "archaeology." They also found a block of grayish-pinkish granite engraved with the writing of three different eras. Years after they returned home, this stone would allow archaeologists and historians to decipher the hieroglyphic script and illuminate a long-lost world.

Many of the men who signed onto the "Commission on Arts and Sciences Attached to the Army of the East," as it was known, were colorful, eccentric characters, brilliant men who kept copious notes. I decided to focus on ten of these *savants*. I selected them for their status in the history of science, for the richness of the materials they left behind, for their relative prominence in the expedition, for their relationships to one another and to the military and political events of the day, and for their involvement in the expedition's final book on Egypt.

Their story is well known in France and Egypt, part of the basic national history lesson in those two countries, but the rest of the world has hardly heard of these men. Egyptologists—scholars and archaeologists who study ancient Egypt—study the *savants'* great book to understand how the archaeological sites have changed since 1800, and historians of science have written scholarly papers for publication in academic journals. Beyond that rarefied community, though, very few people have heard of their adventures and ambitious, beautiful, flawed contribution to East-West understanding, even though their greatest find, the Rosetta Stone—key to the hieroglyphic script—is now a household word, a metaphor for the clues that make decryption possible.

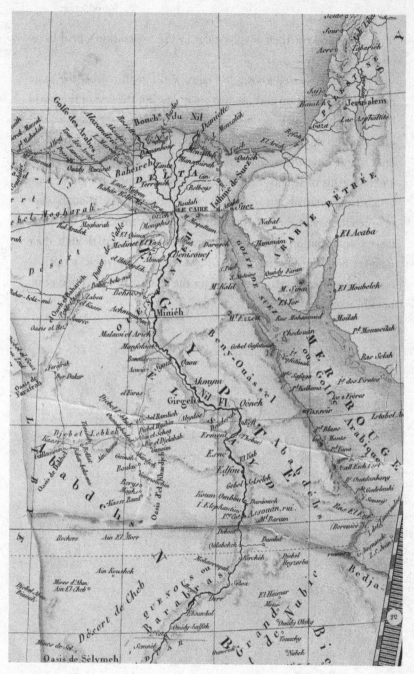

Carte Generale des Etats du Nord de l'Afrique 1828, New York Public Library Collection

CHAPTER 1

THE GENERAL ◿◿◿◿◿◿◿◿

*Europe is but a mole-hill. There never have existed mighty
empires, save in the East, the cradle of all religions, the birth-
place of all metaphysics.*

—Napoleon Bonaparte

*There was nothing more surprising and marvelous than
Napoleon's salon for company: it was like something to lodge
a king.*

—François Bernoyer

Mediterranean Sea, May–June 1798

Departure day dawned warm and sweet, a merry late-May morning on the shores of the Mediterranean. Sunlight winked on the water, the wind was brisk. A military band played martial music, and booming cannon warned stragglers to board without delay. The bay was black with three hundred ships—a sight unlike any the seedy port town of Toulon had ever seen. Vessels were packed so tightly that hulls screeched and sawed against each other maneuvering for open water. Wives, family, friends, and the plain curious milled about on land, sobbing, laughing, waving, struggling to witness this massive, unprecedented embarkation.

The lace-cuffed artist and diplomat Dominique-Vivant Denon surveyed the scene from the deck of one of the ships, and mused in his journal: "Thousands of men leaving their country, their fortunes, their friends, their children, and their wives, almost all of whom knew nothing of the course they were about to steer, nor indeed of anything that concerned their voyage, except that Bonaparte was the leader."

On board the ships, the scene was colorful, chaotic, profane. The infantry wore blue, the hussars (fighting horsemen) sported yellow and red, the dragoons (mounted infantry who fought mainly on foot) draped themselves in scarlet, pink, or orange, depending upon their unit. Horses whinnied and stamped down in the holds, livestock to be butchered in the course of the voyage lowed and grunted from rank pens.

Scattered among the 34,000 land troops and 16,000 sailors and marines crammed into every nook and cranny of this massive fleet, 151 scholars and artists tried to stow their books, instruments, and baggage without disturbing the rough men around them, still busy tying down bulky equipment and hurling oaths as the footing beneath them began to roll. Unlike the soldiers, these civilians had actually *volunteered* for this mad expedition to a secret destination, but like them, they were assigned to ships and berths according to their age and rank and prominence in their respective fields. Professors, inventors, and famous artists like the elegant Denon sailed in relative luxury with officers, while students and young engineers squeezed into fetid holds together with a hundred or more men.

From the start, down in the holds, the youngest scientists—thirty-six students—had the worst of it. As the ships lurched into the rolling open sea, the students clutched hammocks hanging next to, over, and under rows of soldiers. Hemmed in on all sides by human flesh, teenagers who had, a few months before, fought to join the mystery expedition were quickly disabused of their adventure fantasies. Within hours of embarkation—just before dinner, in fact—the rough seas drove most men to their knees with sickness. The soldiers were from the land army, not seafaring men themselves, and they got as seasick as the scientists. Men who weren't subject to *le mal de mer* were soon sickened by the inescapable odor of vomit. Even Napoleon, who had his shipboard bed mounted on wheels to try to alleviate the effects of the swells, spent much of the trip seasick.

It was a queasy and inauspicious start to a voyage the civilians had anticipated with nervous impatience for many days. They had left the comforts of Paris weeks before, and had waited in the port town of Toulon for days. Horses and war matériel clogged the thoroughfares—the streets were filthy, lodging and food were nearly impossible to find. Men slept in stables and on the floors of public buildings. Night and day, tens of thousands of battle-hardened soldiers, returned from the two fronts in France's recent European war, clashed with one another as well as with the scientists. These dirty, rugged, uneducated men would be the scientists' protectors and tormentors for the next three years. They jostled in the fishy streets, drinking and brawling, simultaneously impatient for the sea voyage to begin and annoyed at the impenetrable mystery of the destination. They ridiculed or sneered at the civilian scientists in their frock coats—if they noticed them at all.

The animating spirit behind the enterprise didn't arrive until the night before the expedition was scheduled to sail. On the evening of May 18, Toulon blazed with celebratory lights. The greatest military leader in France, at five-foot-four, a lithe little man who moved like a human panther—"that sulphur-headed Scaramouche," to his royalist detractors—had finally arrived, with his wife, Joséphine. He was twenty-eight years old. A curious combination of romantic and cold-blooded tactician, the young warrior was, at this point in his life, deeply in love with the comparatively more worldly and sexually experienced Creole whom he had married. Leaving Toulon, Napoleon didn't know that before he saw his wife again she would break his heart. With her silk-draped presence nearby, he loped onto a hastily constructed platform and, to roars from the crowd, addressed his soldiers, promising to those who returned from this mystery destination a grant of six acres of land each.

As he spoke, the crowd crackled with a sense of history-making and adventure. The general never had a problem raising massive armies. The French People's Armies had numbered between half and three-quarters of a million men in the preceding years. Before his reign ended, Napoleon would muster a million-man army. François Bernoyer, a tailor and chief of supplies to the army, who would become a prolific chronicler of the expedition, was in the audience. "Everyone was excited by the mystery. Never had so many masts been seen on the sea, nor so many horses and war tools filled the beach. Thus, on the faith of a single man, the elite of the French *savants* and warriors prepared to leave."

Like most Frenchmen in 1798, the nation's scientists worshipped the young general. Napoleon Bonaparte was more than just a celebrated fighting man, he was the nearest thing to a rock star the late eighteenth century would see. Corsican, and a true son of the French Revolution, he had already become a new kind of leader for a new era without priests or kings. When he participated in state affairs, he surrounded himself with secular—not ecclesiastical or royal (that would come later)—pomp. Returning to Paris after taking Italy in 1797, he was fêted with a nonreligious service where divinity-student-turned-diplomat Charles-Maurice de Talleyrand led him to the "Altar of the Fatherland" and a military band serenaded him with a new hymn, not to God, but "to Liberty."

The young general revered science. Later in life he would say that if he had not become a soldier, he would have made world-changing scientific discoveries. He thought of himself as a physicist, a metaphysician even, who would have identified the smallest particles that make up the world, as well as the invisible forces that keep them together. The scientists were flattered by the young general's esteem, and none questioned the hero's bona fides. They had even invited him to join the mathematical section of their exclusive club, the Institute of France. "True victories, the only ones that cause no regret, are those made over ignorance," Napoleon had said at his induction. He was flattered by the honor and throughout his time in Egypt he sometimes signed proclamations, "Napoleon Bonaparte, Member of the Institute of France." Whenever he was in Paris, he never missed a meeting. On January 21, 1798, the sixth anniversary of the beheading of the king, he did not attend the public memorial ceremony as a general, but went inconspicuously among the Institute's members.

The general wasn't so revolutionary as to spurn all the luxuries due a military hero. Setting off from the coast of France, Napoleon sailed on the largest ship in the French fleet, indeed one of the biggest ships in the world. The flagship, formerly called by the Bourbon navy the *Dauphin Royale*, now renamed *L'Orient*, weighed 2,000 tons and carried 120 cannons. *L'Orient* was a surreal amalgam of paranoid war room (packed with gunpowder and men nervously speculating about approaching English ships), floating literary salon, and imperial ballroom. The ship carried not only the general and his considerable entourage—several thousand crew and soldiers—but also a two-month supply of food, including live cattle and sheep, a massive traveling library (with works by Plutarch, popular novels, and textbooks on chemistry and mathematics), and the bulk of the French fleet's firepower. Tons and tons of explosive material crammed into the hold rendered the ship a floating bomb.

Napoleon's own quarters, draped in hundreds of yards of red damask, were outrageously *royal*. "Everyone is talking about Napoleon's room on the *Orient*, which is apparently very luxurious," one of the scholars, the young zoologist Étienne Geoffroy Saint-Hilaire confided in a letter home. The military chronicler Bernoyer was invited to one sybaritic shipboard dinner attended by, he claimed, three hundred people. "When they went to eat, the dinner was good, but they were so tightly squeezed together that they could barely raise their forks to their mouths," he wrote. After dinner, the general hosted operas and concert music on deck by moonlight. At one point in the thirty-day sea journey, military musicians composed music especially for the expedition and played it on the deck of *L'Orient* during a lunar eclipse. "There was nothing more surprising and marvelous than Napoleon's salon for company: it was like something to lodge a king," wrote Bernoyer, who found the scene disturbing and compared it to finding a great lord in the middle of a camp of Spartans. Bernoyer was astonished because although Napoleon was a hero and due tribute, both material and spiritual, he was still just a general, and a revolutionary general at that. He had not previously indulged the royalist tendencies to which he would succumb utterly as emperor.

The scientists, all *citoyens* too, were untroubled by revolutionary scruples about imperial luxury. On the contrary, they were delighted with their frequent invitations to the general's plush suite. After dinner, when Napoleon wasn't too seasick, he gathered his favorite intellectual "mistresses"—the chemists, zoologists, mathematicians, and artists in his *corps de savants*—into nightly salons on deck. These shipboard seminars became the stuff of Napoleonic legend. The young general put the question for the evening according to his whim. *Is there life on other planets? What do dreams mean? What is the ideal form of government?* His living encyclopedia would then have at it, arguing sometimes deep into the night as the black water lapped softly against the sides of the ship.

The men who assembled at these evening salons were as different in their politics as they were in their areas of expertise. The general's favorite was the pugnacious geometer Gaspard Monge. The son of a peddler, Monge, was a revolutionary fanatic. He was short, had a great beaked lump of a nose, and a preference for sidearms and a hip flask. A mathematical genius capable of turning complex spatial relations into numbers on paper, he was deeply enamored of military life. He was always keen to "get a whiff of gunpowder," as one of his biographers put it. Monge invented descriptive geometry, a method of mathematically depicting shapes that allowed mechanical engineers for the first time to invent machines on paper in three dimensions.

Never far from Monge on *L'Orient* was his partner at the helm of the Scientific Commission, the chemist Claude-Louis Berthollet. Where Monge would burst into tears at patriotic displays, the gray-eyed Berthollet appeared unemotional and cold. The general found Berthollet's reserve frankly unappealing, but he respected the chemist's breadth of knowledge and would continue to rely on him for years after the expedition, in matters of both state and science.

Berthollet had been the Commission's chief organizer in Paris. He was tall, quiet, lumbering, apolitical, and painfully shy—Monge's opposite. "Berthollet has a rather ordinary exterior," wrote the younger mathematician and fellow Egypt scholar Joseph Fourier. "He only speaks with the most extreme difficulty, hesitates and repeats himself ten times in the same sentence and seems embarrassed when explaining the least details of an experiment."

Berthollet's reticence belied his genius. He is regarded, along with the chemist Antoine Lavoisier, as one of the founders of modern chemistry, and he pioneered the field of physical chemistry. Before, during, and after his time in Egypt, Berthollet invented new ways to make gunpowder, steel, dyes, bleach, even rudimentary sugar substitutes.

Monge and Berthollet had shared adventures with the young general before, serving him in some quite nonscientific capacities. In Italy, Napoleon put them in charge of organizing and carting off vast quantities of medieval, Renaissance, and classical artistic loot. When Napoleon set off for Egypt, he regarded Monge and Berthollet not just as scientists but as trusted members of his inner circle. The soldiers, familiar with the close bond between the two men, took to calling them "Mongeberthollet" and occasionally argued over which was which.

The wittiest and most sophisticated participant in Napoleon's shipboard salons had no scientific training at all, but he could hold his own in any conversational circle. Dominique-Vivant Denon, at fifty-four the oldest member of the civilian group, was a polymath, the embodiment of the quintessential eighteenth-century European. Artist, wit, diplomat, writer, and explorer, Denon was not invited to join the mystery voyage at first. When he learned of it, he made use of every connection he had, including Napoleon's wife, to get himself enlisted.

Bright-eyed and lively, Denon, though the oldest of the savants, possessed abundant youthful curiosity, a quality that not only sustained him through the hardships to come, but made him the most famous of all the scientists in his lifetime. A humanist by nature and a diplomat by profession, Denon had lived the Revolution from both sides—as a hunted noble associate of the royals (he was a prerevolutionary baron) and as a successful and wily political Republican avoiding the Terror. In the service of the Bourbons, he had been France's representative in, among other royal courts, Florence, where he was renowned for his elegance and charm, and his way with women. He was also a skilled fine artist and he rehabilitated himself after the Revolution with the help of the painter Jacques-Louis David, who put him to work designing Roman-style togas as the official French Republican government costume. In addition to art and diplomacy, Denon was also a writer, and even a pornographer (one collection of his etchings was titled *L'Oeuvre priapique*). Charming, fluent in languages, and skilled in seduction, he was a favorite of the Parisian ladies including Joséphine Bonaparte.

Napoleon's shipboard salons also included younger scientists who gazed at the general in awe as the talk floated into the night air and the military band played Haydn in the moonlight. The sensitive, hero-worshipping young zoologist Étienne Geoffroy Saint-Hilaire, just twenty-eight, was already a zookeeper and professor, but he was about to become a prolific letter-writer and highly unlikely heroic world explorer. Unmarried, childless, bespectacled, pudgy, and soft, with full lips and slightly mournful eyes, he was relatively physically fit, and highly esteemed by his older colleagues. Emotional and high-strung at the best of times, the young zoologist never failed to thrill at his proximity to Napoleon, a man his own age.

For Napoleon's amusement, this select group of civilians debated politics and discussed dreams, extraterrestrial life, and many more esoteric subjects the young general posed to them. One topic that never got aired at the salons, however, was their secret destination and why they were headed there in the first place. As they sipped liqueur on deck, only Napoleon's inner circle knew—or thought they did—exactly where they were sailing. Even if they had discussed it, it is unlikely that the *savants* would have agreed on all the personal and political reasons behind the mission.

England, not Egypt, had been the key to mobilizing the soldiers. The French government called the 50,000 men leaving from Toulon an "Army of England" heading off to face the British. Colonial competition between England and France had grown more intense throughout the 1700s. The two countries had fought each other all over the globe for distant turf—in America, the Caribbean, the South Pacific. The French Revolution had only heightened the tension between the long-time rivals.

The Revolution, culminating in the public beheading of a king, had so horrified and frightened the European monarchies that they joined forces to put a stop to the infection before it spread to their kingdoms. France had been at war almost continuously in Europe for nearly a decade, spreading the gospel of *liberté* and fending off royalists in Austria, Prussia, Holland, Spain, and England. By the late 1790s, thanks in large part to the military successes of young General Bonaparte, the French had negotiated peace with Spain, Holland, Prussia, and even the mighty Austrian Empire. By 1798, only England remained at war with France.

The invasion of Egypt was unprovoked and probably could not have been attempted by anyone but Napoleon. In 1798, his military feats were already so great, and his legend waxing so bright, that whatever he attempted seemed to be dusted with gold. The timing also had much to do with the condition of the French government. France's rulers at this point were five men who formed a governing body called the Directory and who shared executive power until Napoleon, post-Egypt, established his one-man Consulate. The Directory was the last in a series of revolutionary-era governmental bodies, and it operated as a nominal parliamentary democracy. The Terror-scarred French had by this time abandoned the revolutionary goal of universal suffrage. The National Assembly and Estates General, legislative bodies that had opened the revolutionary era in France with the promise of a radical form of populism, were history. The five directors were elected from a list of 500 men by a center-right parliamentary body called the Council of Ancients, who were themselves selected by another parliamentary body, the left-leaning Council of Five Hundred. The directors were unpopular, they ignored the Constitution, and had prolonged their power (and tried to replenish the bankrupt national treasury) through war. They had little to lose and much to gain by trying to seize a new colony and by agreeing to send a popular leader and potential rival out of the country to do it.

The French government had some specific instructions for the young general about what he was to accomplish in Egypt: "The general in chief of the Army of the Orient will seize Egypt; he will chase the English from all their possessions in the Orient. He will then cut the Isthmus of Suez and take all necessary measures in order to assure the free and exclusive possession of the Red Sea for the French Republic."

The notion of Egypt as a prize in play gained credence in the late 1700s as the vast Turkish Ottoman Empire—which, besides Egypt, included a swath of North Africa stretching west across the Sahara through Algeria, as well as (in Europe) Greece, Crete, and Macedonia, and (in Asia) all the land along the Mediterranean coast from Turkey to Egypt then known as Syria—was losing its power. Many European governments believed the entire Ottoman Empire would soon be up for grabs. French leaders, who had a longstanding alliance with the Turks, knew that any of the contemporary rulers—Russia's Catherine II, Austria's Maria Theresa, Frederick II of Prussia, in addition to England's King George—could decide to make the first move.

Napoleon wasn't thinking about the Turks when he set his sights on Egypt. He was thinking about England. He would have liked to confront England directly, but even he admitted that was impossible. Taking Egypt would sever a major English-Indian trade route, though, and weaken British imperial aspirations.

Napoleon also shared with future Europeans what the historian Christopher Herold called "the great Victorian folly"—an obsession with the Orient that would become a dominant political sentiment of the coming age. As a young man, he had volunteered to go to Istanbul and teach the Ottoman military some new tactics. (He never went.) Young Napoleon was so intrigued by Egypt that he admitted in private letters that he would happily trade all of Italy itself—his greatest triumph so far—for firmer control over tiny islands like Corfu and Cephalonia, footholds in the Mediterranean Sea from which to better fortify a French presence in Egypt and the Levantine coast.

Egypt was a tempting colonial prize for the French because it was just across the Mediterranean. It was also the gateway to Africa and Asia, and it was potentially, if not actually, rich. The Nile valley could provide sugarcane, flax, indigo, wheat, rice. Anyone who seized Egypt controlled the natural resources not only of the desert (alkali, mainly) and the Nile, but also the gold, timber, and other raw materials from deeper Africa and Asia that crossed Egypt via caravan. Equally important, control of Egypt meant a geographic foothold near Asia, and a step closer to that Oriental jewel, India.

The colonial motive alone hardly justified the French sending men, matériel, and their greatest military leader away in the year 1798, but taking Egypt was a way for the French to engage their greatest European foe without challenging them directly at sea. Following the loss of his American colonies, King George III preferred to avoid another war, but when the French guillotined Louis XVI, he was all for fighting "that most savage and unprincipled nation," and the two nations' rivalry grew more heated. It now involved such fundamental differences as monarchy versus revolutionary republic and religion versus atheism.

Egypt, on the other hand, was reputed to be easy to conquer. Both sides agreed that a French Egypt would put a damper on British colonial expansion, and be a sort of triumph by proxy. The French consul in Egypt had predicted that a French victory there would register as "a conquest taken from the English." A prominent British merchant familiar with the region warned his countrymen that if the French invaded Egypt they would possess "the Master Key to all the trading nations of the Earth."

France's ties to Egypt were already stronger than England's. Fifty French merchants operated in Cairo and Alexandria; the French had long-established consulates at Alexandria, Rosetta, and Cairo. Both countries bordered the Mediterranean. While the French were most numerous, a tiny community of European traders, including Englishmen, had operated in Egypt for centuries. The European merchants were rivals, but in 1798 they shared a common problem: none was welcome in Egypt. Egypt's tax-sponging Mameluke leaders harassed them for money—a problem the British had recently tried to remedy by negotiating with the Mameluke beys, something the French, allied with Turkey, could not do. Moreover, outside the cities, nomadic bandits hindered commercial movement across long and waterless stretches.

One unpleasant and relatively recent incident was well known in France when Napoleon and his fleet sailed. In 1779, a gang of Arabs had stopped a caravan of French merchants traveling from Suez to Cairo with their wares in the desert. "They stripped them naked," one French writer recounted, and sent them off into the desert to meet their fates. The sole survivor, a Monsieur Saint-Germain, barely made it back to civilization "without any shade but a thorn bush and forced to drink his own urine. His body was one entire wound, his breath cadaverous, when he arrived in Cairo."

Contemporary conditions notwithstanding, France in the 1790s was already enamored of ancient Egyptian iconography. They imagined the civilization of the pharaohs as a kind of purer, natural society that preceded corrupt kings and Catholics. In addition to employing Egyptian motifs in revolutionary memorials and architecture, the new government had signaled its intentions toward the modern Orient by establishing a public school in the national library to teach Arabic, Turkish, and Persian languages. Almost all of Napoleon's translators in Egypt were products of this program.

On the sea journey, the scholars could and occasionally did provide instruction and general amusement to the rank and file outside Napoleon's private salons. Astronomers shared their telescopes as they passed Sicily, so common soldiers could peer at smoking Mount Etna and laugh at the nervous crowds of Sicilians gathered on shore watching the ominous French flotilla. Geoffroy Saint-Hilaire conducted public experiments with galvanism (the production of electricity via chemical reactions) and dissected a large shark for the sailors' edification. The same shark, while still alive, flapped its tail, knocking over five sailors and spraying Napoleon himself with blood. Geoffroy Saint-Hilaire also toppled into the sea while changing boats on one of his errands and nearly drowned, a spectacle that amused the soldiers at least as much as did the shark's internal organs.

Geoffroy Saint-Hilaire enjoyed the sea journey, and found nothing to complain about, especially in the food department. He was, though, amazed at the rate with which men like himself seemed to fall overboard. He wrote to his mentor in Paris, the naturalist Georges Cuvier, that one could "not imagine how dangerous and disagreeable the navigation of a fleet could be."

Mortal dread shadowed the midnight salons, daytime shipboard amusements, and seasickness. Every time a telescope sighted a speck on the horizon, Napoleon and his generals envisioned the English commander Horatio Nelson bearing down on them with the terrible might of the English navy. And as the bulk of the expedition's explosive material was imprudently crammed onto just one vessel, L'Orient, those who sailed on Napoleon's ship were more tense than most. At the slightest provocation, including on some occasions, French ships that had drifted away from the flotilla, soldiers scrambled to their guns, with turmoil and racket all around. The scientists were put to work lugging fire hoses during these panicky exercises.

After twenty-two days at sea, the French fleet reached the island nation of Malta, where troops easily won control from the outnumbered, outgunned Knights of Malta. Napoleon chose the island to be a strategic foothold for French ships in the Mediterranean once Egypt was taken. The Knights, a religious-military order that dated back to the Crusades, surrendered with a little negotiating help from one of the older members of the scientific corps, the fifty-year-old geologist Déodat de Gratet de Dolomieu. Dolomieu, whose name is immortalized in the Dolomite Mountains, where he did much of his research, also happened to be a Knight of Malta, since his parents had entered him in the order at the age of two. Napoleon insisted the geologist negotiate the Knights' surrender, and Dolomieu reluctantly acquiesced, a decision that came back to haunt him soon after. The Knights got their revenge on Brother Dolomieu within a year, capturing and jailing him. The geologist never forgave Napoleon for assigning him to betray the order.

The French—liberators first, of course—freed the slaves held by the Maltese, mostly Turks and Africans. They then loaded their ships with loot—coins, art, silver plate, and bejeweled weapons, five centuries of booty, including gold valued at five million francs. Malta was, however, a mere way station for stocking up on treasure to help bankroll the expedition. Once the troops were back on board the ships, Napoleon announced their true destination.

Despite rumors and suspicions, most soldiers and many members of the scientific expedition had been in the dark regarding their goal, even after weeks at sea. On June 22, Napoleon ended all speculation. He ordered his captains to read a proclamation aloud on all the boats, announcing that they were in fact en route to Egypt. "Soldiers!" it began. "You are going to begin a conquest of which the effects on civilization and world commerce are incalculable! You will give the English a most sensible blow, which will be followed up with their destruction."

The young general's devoted fans, including all the civilians, were untroubled by the fact that Napoleon was leading 34,000 soldiers into Egypt at a critical moment for France. Such was their faith in the living myth of the first great postrevolutionary French hero that the scientists traveling with him, like the soldiers, never yet doubted his military instincts.

Now the voyagers had an end point to ponder. What they and their general knew of it was a tapestry of ancient myth and European travelogue, woven through with misinterpretation. Before he left Italy, Napoleon snatched from the library of Milan all the books about Egypt and the Middle East. These texts together contained the bulk of European understanding of Egypt circa 1800.

The classical texts laid the groundwork. The oldest Egyptian travelogue came from the fifth-century B.C. Greek writer Herodotus, who wrote about pharaonic Egypt when it was already more than 2,000 years old. The ancient civilization was in decline when he arrived. Herodotus catalogued Egypt's flora and fauna and geography, and the French relied on his accounts as a basis for what to expect.

Many writers had speculated on Egypt over the ages. Plato thought pharaonic Egypt was the primordial source of human culture, and that its people existed in a golden age of civilization. Arabs in the Middle Ages and Renaissance European scholars maintained the notion that the Egyptians had access to lost esoteric knowledge.

A few intrepid European travelers had recently contributed detailed, if delirious, observations on the modern Islamic nation. All together, classical and contemporary accounts had produced an image that was faithful to reality—desert, ancient monuments, and Muslims—but also fantastic—medicinal mummies and lost magical arts, awash in gold, silks, and perfumes, ruled by men who lived and died by the ruby-crusted saber.

As Napoleon sailed toward Alexandria, two contemporary accounts of Egypt vied in his mind. Travel-writer Claude-Étienne Savary returned from Egypt in the 1780s with a favorable if condescending view, and reported finding "the tranquil garden promised by Mohammed" in the delta city of Rosetta. "Here the Turk [some Europeans tended to call all residents of the massive Ottoman Empire Turks] smokes all day pipes of jasmine and amber, in a fragrant place, thinking little, without ambition, activity, the soul of all our talents, is unknown to him. He peacefully enjoys what nature offers, what each day presents to him, without worrying about the next day." Savary's garden of sensual delights included naked bathing beauties on the banks of the Nile.

Napoleon had also read and met with another travel-writer, the ponderously named Constantin-François de Chassebouef de Volney. Volney was more critical, although his dour text was fantastically illustrated by the traveling artist Louis-François Cassas, who depicted an Oriental nirvana of reconstructed ruins and sylphs bathed in shadows of obelisk and palm.

Volney had visited Egypt in the 1780s, and he disabused his French audience of their silk and spice fantasies. The author's bouts of disease and accounts of inhabitants harassing French merchants produced a less pleasant image of the land of the pharaohs.

In the winter months before he left for Egypt, Napoleon had long conversations with Volney, who described French merchants living under conditions close to house arrest. "The merchants remain at Cairo at the peril of their lives and fortunes," Volney reported, giving the general another pretext for invasion. "They are shut up at a confined place, they live among themselves with scarcely any external communication. They go out as little as possible in order to avoid the insults of the common people who hate the very name of the Franks."

Volney's jaundiced view did not eradicate the general's Oriental fantasy. Presumably, Napoleon was prepared to expect the worst, but as his ship rocked on the Mediterranean at night, the general nurtured a notion of Egypt as unreal as any other European's. He had been imagining the Orient since boyhood, when he wrote a short story set in the Ottoman Empire, involving French soldiers and the daring rescue of slave girls from a harem. In young adulthood, savoring life as a national hero, he began thinking of himself as a world conqueror in the style of Alexander the Great. Alexander's greatest triumph, of course, was the conquest of the East.

Napoleon was not alone in his fantasies. All the French expedition members had different variations on that same exotic theme in their minds. Even the scientists, bored to distraction during their final weeks at sea, indulged private notions of a land of rich soil, colossal ruins, classical columns, and even libraries of ancient knowledge. The Egypt that intrigued them was a land of classical ruins, not the ragged edge of the faltering Ottoman Empire, inhabited by Muslim Arabs and Africans. The vision shimmered on the horizon of their minds as they sailed those last days due southeast from Malta.

It was immediately dispelled at the sight of the fabled port city of Alexandria.

CHAPTER 2

THE GEOMETER
AND THE CHEMIST

The first city we shall arrive at was built by Alexander, and every step we take we shall meet with objects capable of exciting our emulation.

— Napoleon Bonaparte, in his first proclamation to
the troops sailing toward Egypt, June 22, 1798

Monge loved me as one loves a mistress.

— Napoleon, recalling his geometer's
umconditional adulation

Alexandria, July 1798

On July 1, at four in the afternoon, first one, then another, then a chorus of voices from three hundred ships called: "Land!" The first glimpse of Egypt filled each man with joy, expectation, and dread. Men clamored for the telescopes, seeking a glimpse of the ancient capital of classical learning and luxury. They squinted into the lenses. The afternoon sun glittered on the water, blinding them at first, confusing their eyes. Was it possible that, for their first sight of Egypt, the fabled origin of civilization, this was all?

Water. Sand. And a desiccated town set against a great white void.

The French didn't have time to indulge their disillusionment, because they were in mortal danger. The long beach at Alexandria is curved, with narrow strips of land extending into the sea. Between these encircling claws of earth, the Mediterranean Sea boils and thrashes, disintegrating into white spray on huge, barely submerged rocks. Even today, fishermen in small boats who know the waters well struggle for control in the turbulence, barely dodging the underwater granite boulders with jagged points that are visible, breaking the surface, only occasionally, between swells. The bay had always been treacherous, which is why the ancients built a lighthouse here called Pharos, one of the seven wonders of the world, to help sailors make their way. The seafloor is still littered with the wreckage of centuries of sailing ships, including French vessels from two hundred years ago.

Everyone—sailor, soldier, general, and scientist—felt a certain unease on first sighting Egypt. It wasn't just the rough water and looming storm, or the certainty of a coming battle with the guardians of Egypt, the equestrian Mameluke fighters. For the soldiers, it was the beginning of disappointment and worse. Seeing clusters of ruined buildings against the horizon, site of one of antiquity's most fabled cities, a soldier standing next to Denon joked to a mate that there lay the six acres of land Napoleon had promised each man back in Toulon.

The scientists, however, could barely contain their excitement. What they saw through the spyglass was the shimmering shore of Africa, a great, undiscovered continent teeming with new flora, fauna, and people.

It had taken just two months in the winter of 1798 for Berthollet to enlist 150 men for the scientific corps. The quiet chemist took charge of the overall organizing since his cohort, Monge, was still in Italy on business for Napoleon (including stealing the Vatican's printing press with its Arabic letters). Working with the leaders of the two main universities in Paris, Berthollet personally solicited many of the scholars. His task was surprisingly easy. Students especially clamored to join up and follow the hero-warrior, and Berthollet was spoiled for choice. He chose the cream of the crop, the smartest boys and their most inventive teachers.

A few turned him down. France's foremost naturalist, Georges Cuvier, declined the invitation, judging—probably correctly—that, professionally, he would benefit more by staying in France. "I was at the center of science in the midst of the finest collection," he later explained, "and I was sure to do my best work there, more coherent, more systematic, and containing more important discoveries than in the most fruitful voyage." Cuvier sent a young botanist named Savigny in his place.

The *savants* who answered the general's call welcomed the mysterious expedition as both an adventure and a professional opportunity. Paris was the national center of the sciences, but it offered little for young, ambitious men in 1798. The nation was in near ruins after the bloody explosion of liberty, equality, and fraternity earlier in the decade. Economic depression, food shortages, worthless currency, and war were crushing the nation. The chaotic, weak government could promise only more of the same. Bread was still rationed. And though the streets were no longer sticky with gore, order was barely restored. Bandits terrorized rural travelers, and street gangs clashed in the city.

Yet intellectually postrevolutionary France was still vital, if quite altered. Gone were the *ancien régime* salons where witty banter was raised to an art form. Members of the aristocratic intelligentsia had been killed or had fled, and those who remained or returned were changed, nihilistic remnants of the *salonnards* of two decades before. Rather than meet in drawing rooms to discuss philosophy, they gathered in ballrooms at debauched fêtes, where low and high, prostitute and aristocrat, mingled with a single goal in mind: forgetting.

The Revolution temporarily disrupted education and altered the manner of teaching in French schools, indelibly marking the careers of every young scholar on the Egyptian expedition. Before the Revolution, Catholic clerics ran most of France's schools. Bookish boys of the middle class mostly aimed for a career in the priesthood. The Revolution closed that professional option along with the Catholic schools. With the priests and monks banished or demoted, and plans for a new secular education system not yet realized, the schools barely functioned. Government primary schools that were open, though, taught a new catechism to the nation's children: "What is the soul? I know nothing of it. What is God? I don't know."

France, having expelled priests and nationalized Church lands, was officially godless. New secular celebrations, including the macabre annual commemoration of the beheading of the king, had replaced the old saints' days. With the nation worshipping at the altar of Reason, scientists served as spiritual guides to men like Napoleon. The new Republic wanted men of thought *and* action, questing and inventing. The era of professionalized science as we know it was still decades off, but the dilettante eighteenth-century man of science—the aristocratic amateur—was fast becoming a rejected relic of the *ancien régime*. Science was also becoming more sophisticated, and scientific methods, while still crude, were beginning to look modern with their focus on rigorous experimentation.

· The Revolution had left Paris a shambles, yet France remained an international leader in the sciences. Revolutionary scientists had already invented a new national measuring system, the metric system, to replace the inexact, feudal system of weights and measures associated with royalty. The Revolutionaries had also redefined time itself, discarding the old Julian calendar, with its religious feast days, and replacing it with something both more pagan and more modern, based on the natural rhythms of nature and agriculture. Now the new year began at harvest time, in the newly named month of *vendémiaire* (roughly, vintage), which was followed by the cooler months of *brumaire*, *frimaire* and *nivôse* (roughly, mist, frost, and snow). The year ended in late summer, in the month of *fructidor* (roughly, fruit). Throughout the years the scientists spent in Egypt, these freshly named months would be the measure of their days.

The two major institutions of higher learning in 1798 Paris—the

Museum of Natural History for the biological sciences and the École Polytechnique for mathematicians, physicists, and engineers (the world's first engineering school)—were beacons of secular promise. The newly named Institute of France, a professional academy where the nation's greatest thinkers were inducted, had existed before the Revolution under a different name as a royal science academy (one of four royal academies of arts and learning), and still maintained strict standards, dedicated to skepticism, classification, and demonstrable proof.

The scientists who enlisted were helping to glorify France, and their peers eagerly supported them. In war-ravaged, depressed Paris, Berthollet assembled a massive traveling library and requisitioned an enormous collection of tools, including the entire laboratory apparatus of Paris's preeminent school of science, the École Polytechnique. The instrument-makers of Paris yielded up all the glass, steel, brass, and wood devices in their workshops to the expedition. The scientific gear eventually occupied the hold of an entire ship: crates of scalpels, glass jars, alcohol, microscopes, and magnifying glasses for the naturalists; paints, pencils, brushes, and paper for the artists and engineers; telescopes for the astronomers; tons of measuring and surveying equipment for the mapmakers.

Neither students nor mentors were deterred by the fact that the actual destination was an official secret. The professionals, who would earn a stipend, were dazzled by the prominent organizers and by the promised proximity to the general. They believed their positions would be protected in their absence (although as time wore on they felt less secure on that score). The students, with less at stake, chased the dream of travel and adventure.

Historians of the expedition believe that only about 150 of the 50,000 men who set off from Toulon were officially informed of the destination. Most soldiers and some lower-ranking civilians remained in doubt about their goal until Napoleon's announcement on the journey's final leg.

The mobilization of France's intellectual elite for this secret expedition was not entirely secret, and most of the civilians strongly suspected the final destination before they signed on. On April first, more than a month before the flotilla set sail from Toulon, the Parisian journal *Le Moniteur* reported that the most distinguished men of science were participating in some kind of military and scientific expedition to a "destination in another part of the world." A few weeks later, the newspaper actually named the rumored destination, reporting, "They speak of Egypt." At the Museum of Natural History, the destination was no secret at all, and Cuvier was openly calling Geoffroy Saint-Hilaire and Savigny "our Egyptians." Geoffroy Saint-Hilaire was sure of the destination before he left, because of the books being gathered for the Commission. "When he left Paris, he had prepared himself for the scientific exploration of Syria and Egypt," his son Isidore wrote in a biography.

As the ships tossed in the increasingly stormy Alexandria Bay, among the *savants*, only Monge, mad for the glory of France and eager to delight his hero, was for assaulting Alexandria immediately. The geometer had to be restrained from leading a charge. For the rest of the scholars, the prospect of an imminent battle had a sobering effect. As dusk fell and the fleet of ships bobbed in the bay, awaiting orders, the artist Denon peered through a telescope and contemplated the outlines of the decayed town, comparing it to the great cities of classical lore. Among the palm trees of the few gardens, he saw the silhouettes of minarets and Pompey's Column, ancient landmark to seafarers and nomads in the surrounding desert. "My imagination went back to the past," he wrote later. "I saw art triumph over nature; the genius of Alexandria employ the hands of commerce, to lay, on a barren coast the foundations of a magnificent city; the Ptolemies invite the arts and sciences and collect that library which it took barbarism so many years to consume." In rapture now, he concluded: "It was there, said I, thinking of Cleopatra, of Caesar, and of Anthony, that the empire of glory was sacrificed to the empire of voluptuousness."

Black clouds blew in just as night fell. At midnight, in slashing rain and with only lightning piercing the darkness, the French consul at Alexandria rowed out to the fleet with some very bad news: the British had been in the area recently, just three days prior, and might very well be returning at any moment.

The report convinced Napoleon not to waste time waiting for the storm to pass. He ordered his men to disembark in what was now a churning sea. Thousands clambered down ladders and ropes swaying wildly in the dark, amid peals of thunder and flashes of lightning, into rowboats pounded by walls of water. Many boats overturned, pitching men into the water, and the cries of the lost continued throughout the night. Anyone who managed to crawl into a rowboat was violently seasick.

After four hours in the swells, rowboats packed with drenched and sick soldiers reached shore. The sea was so rough that even the large boats couldn't put down anchors. Denon watched the action from a perch in one of the frigates. "The little boats received one by one and at random those who descended from the vessels; when they were filled, the waves appeared every instant on the point of swallowing them, or at the mercy of the wind they were forced upon each other."

Bonaparte spared his living encyclopedia this dire adventure. He herded them together onto one ship and left them floating in the bay for days, waiting for the all-clear. Disastrously for the scientists, though, the boat carrying all the instruments Berthollet had so carefully assembled in Paris foundered in the bay that night. Telescopes, microscopes and scalpels, botanists' drawing paper, alcohol for the naturalists' specimens, surveyors' equipment—all sank to the bottom of the bay. This accident, so crucial to the civilians, presaged far worse naval disasters to come, as the French navy's relative ineptitude would soon exact greater costs in lives and equipment.

The nightmare ended with the dawn. As the sky turned pink, the lucky surviving troops were drying themselves on shore, or still crawling out of calmer seas. A few Arabs in their white robes watched under the already burning sun. Hundreds of brass casks winked in the sunlight, a multitude of dripping horses shied and whinnied on the beach; men heaped tackle and artillery into small hills. The Arabs seemed more amused by than afraid of the Franks (the Egyptians had called all Europeans Franks since the Crusades), who were decked out in their heavy, many-colored clothes. Some were quite dandy. One dashing cavalry officer emerged from the sea with intact pink trousers, plumes, ruffles, and a yellow jacket. Most alarming to the Arabs were the chasseurs (light cavalry), clad in green. To Muslims, green was, and is, a holy color, worn exclusively by true descendants of the Prophet. Clearly, these cursing, sweating Franks were no kin to Muhammad.

What the French saw on the beach were Arabs who seemed utterly unimpressed by—in fact, barely interested in—them. After glancing at the invading force, the Arabs went back about their business, which at that hour chiefly involved performing their morning religious rites. There on the sand, the French first witnessed the otherworldly spirituality of true adherents to Islam. The tailor Bernoyer, noting the Arabs' lack of concern at the sudden arrival of cannon, men, horses, and artillery, was flabbergasted. While more than one thousand men were transporting tons of material from the boats to the beach, Bernoyer noted that the Arabs merely walked past them, to the sea's edge, washed their faces and their entire bodies, spread out their clothes on the sand to kneel upon and turned to the East to tranquilly pray, then left, without seeming surprised by the invaders.

The *savants*, and the French generally, were always astonished by what they interpreted as the blasé attitude of the Egyptians. They were baffled by, and never wholly understood, the Arabs' lack of interest in them. To the end of their stay, they debated whether the Egyptians were stupid, possessed of a lack of curiosity, or merely "inscrutable."

On their first day in Egypt, the French army erected huts along the beach and set up headquarters in the white, medieval seawall fort of Qaitbey, built by the Mameluke Sultan Qaitbey in 1480 on the site of the old Alexandria lighthouse. Parts of the lighthouse foundation remain underwater to the present day.

Many competing powers had fought over and controlled Egypt between the end of the Pharaonic period and the arrival of the French. Greeks and Romans ruled it for centuries, followed by six hundred years of Arab control, then Mameluke control, and, finally, nominal Ottoman rule. The Mamelukes built the fort at Alexandria to repel a sea attack on Egypt by the Ottoman Turks.

A white, curving labyrinth of stone tunnels and hallways, Qaitbey's walls were fifty feet high and pierced with gunholes. From this fastness, Napoleon scanned the sea. He wasn't worried about whatever was massing behind him in Egypt's interior; his infantry could handle equestrian warriors. He was obsessed with English sailing ships and European cannon power. He saw only the French fleet bobbing out on the water, but he feared the British might sail back into Abukir Bay at any moment. To risk sailing into Cairo via the mouth of the Nile was courting danger. Weighing the two possibilities against each other—a disastrous sea confrontation with the English, or a tortuous desert march to Cairo—he chose honorable death by sunstroke over dishonorable defeat. He ordered his army to march five days across the Sahara to the capital of Egypt.

In a rush, the general didn't bother with minor details such as canteens and water. He did, however, see to it that his men were supplied with French translations of the Koran. He spared his scientists the arduous march, but he invited Monge and Berthollet to ride horseback with him at the head of the army. As usual, the scholarly duo could not decline the honor.

Led by their general, his enthusiastic geometer, and the quiet chemist, 30,000 men clad in Alpine wool uniforms tramped out into a sea of burning sand on July 8, 1798. Soon after leaving Alexandria, heat penetrated the soles of the soldiers' boots. Before long, their heads were on fire inside their hats, and men started collapsing. They began to understand something they could not have comprehended in Europe: that the sun alone can kill a man.

Almost worse, another phenomenon also unknown in Europe tormented those who remained upright and moving. At the far edge of the broiling world, they discerned shimmering lakes, palm trees, even houses. Water! They quickened their pace, staggered on, only to see paradise shrivel and recede at their approach.

On the first day, they "breathed a burning vapor" and prayed for the sun to set, wrote Bernoyer, who accompanied the marchers. When dusk arrived, there would be no rest, because Napoleon ordered the army to continue marching to take advantage of the relative cool of night. Dawn brought a day no better than the first. "I walked among the soldiers, my knees trembling with feebleness, a thick froth on my lips, and with my throat tightening and stopping my breath," Bernoyer wrote. "In this deplorable state, we continued to make some extreme efforts to reach a large area of water that seemed to be in front of us. But at every step, this precious water fled, always remaining at the same distance. It gave us an ever-renewing hope that at the same time duped us. So despair took possession of our minds and the march slowed down."

As the hours passed, stupefied soldiers stopped hearing the complaints and moans of their comrades, or even noticing those who fell to the ground next to them, dying of heatstroke. "I saw them fall, almost at my feet, without being moved, so much did personal sufferings close one's heart to all feelings of humanity," Bernoyer wrote.

The army's chief surgeon, Dominique Larrey, accompanied the first wave of soldiers into the desert. "Marching on foot, under the rays of the burning sun, over sand which was even hotter, across immense plains of a frightful barrenness, where it was only occasionally possible to find a few stagnant pits of foul water—so muddy as to be almost solid—even the strongest of soldiers devoured by thirst and overcome by heat, succumbed under the weight of their arms. The vista of distant lakes, an effect of the mirage, seemed to promise an end to our woes, but only served to aggravate our distress and to produce the extreme form of collapse and prostration that I saw in many of our brave fellows. Summoned to them, too late to be of any use, I saw many perish of exhaustion. Such a mode of death seemed peaceful and even pleasant, for during the last moments of his life, one of them told me he was 'comfortable beyond words.' "

The ordeal lasted five days. Untold hundreds died of heatstroke or committed suicide.

Geometer Monge and chemist Berthollet, on horseback and riding ahead with Napoleon, were consummate men of science. They looked forward, or, in Berthollet's case, down at the alkaline soil, and not behind at the collapsing men. The two men could never resist taking a closer look at the rocks, the animal bones, the antiquities jutting out of the sand. They kept stopping to inspect, Napoleon eagerly joining them. As the desert trek continued, thirsty men started blowing out their brains in desperation. The soldiers finally turned their rage on the *savants*, spewing threats and oaths on them personally.

Berthollet was unmoved. He collected samples of naturally occurring natron, a desert substance the ancients had used for drying mummies. He noted with interest the high concentration of sodium carbonate (soda) by the dry Lake Natron on the edge of the desert. As he rode along, he began calculating just how, under the prevailing physical conditions, sodium chloride in the upper layer of soil had reacted with the calcium carbonate from nearby limestone hills. He began theorizing that chemical affinities are affected by physical conditions, as in the Egyptian desert, for example, where the natural heat and high concentration of calcium carbonate together created soda. Years later, back in France, this discovery was hailed as groundbreaking in the field of physical chemistry.

Riding beside him, Monge grew ever more fascinated by the optical trick on the horizon. During bivouacs in the burning wasteland, he dismounted, opened his notebook, studied the "water" shimmering on the horizon, and sketched out a diagram of how dense heated air on the earth's surface could create optical distortions. He presented his theory of the *mirage*—a word rooted in the French verb *mirer*, to look at—to his fellow French scientists in Cairo a few weeks later. His hypothesis is remarkably close to the way modern optics explains the phenomenon.

The fifty-two-year-old mathematician was three decades older than most of the soldiers and *savants*, but his zeal belied his years. Early in the Egyptian campaign, he enthused in a letter to a friend in France: "Here I am, transformed into an argonaut! This is another one of those miracles produced by our new Jason who is going to carry the torch of enlightenment to a country which for such a long time has remained in darkness."

The son of a peddler, Monge's success would have been unlikely anywhere in eighteenth-century Europe except in postrevolutionary France. Born in 1746, he displayed an early genius for spatial relations. At fourteen, he constructed a fire engine, at sixteen he made a map of his small town, Beaune, with surveying instruments he made by hand. The precision of the map attracted the notice of the military, which asked his father to allow the boy to enter military school. Monge's father agreed, unaware that his lowly birth would limit his chances for advancement within the royal system.

At school, Monge cheerfully did the dirty practical jobs. One of the most time-consuming parts of a military education at that time was working out mathematical models of designs to assure that no side of a fort was exposed to direct fire. One day Monge presented a plan for building fortifications without performing any tedious calculations first. The supervisor at first refused to look at it, and chastised the teen for not doing the math. Monge insisted he had figured it out another way. When they checked his work and found it accurate, his amazed professors realized their student had invented an entirely new way of applying mathematics to the physical world. Monge was sworn not to share it, and he was not allowed to teach it publicly in Paris for another twenty years.

Monge's invention, known as descriptive geometry, is simply a method for representing figures in three dimensions on one plane, by making two projections of the figure—a top view, or "plan," and a side view, or "elevation." This simple invention revolutionized mechanical design.

Monge's discovery became one of the fundamentals of an education at the École Polytechnique, the postrevolutionary school in Paris he helped found. Monge had great faith in the implications of his work, writing at one point that "Descriptive geometry is a language necessary for ingenious persons." Monge felt that this language "had a transcendent quality" in that it put human faculties in contact with the mathematical structure of external reality, as one science historian has written.

In person, Monge was no cold pedagogue but a passionate fighter and zealous revolutionary partisan. As a poor youth, he overheard a nobleman slandering a young widow who had rejected him. Shoving his way through the laughing crowd, he punched the cad in the nose. The lady in question later became Madame Monge.

During the Revolution and subsequent wars, Monge oversaw the stockpiling of gunpowder and weapons when the national arsenal was empty. With Berthollet, he facilitated the speedy fabrication of huge amounts of saltpeter, a primary ingredient in gunpowder, and persuaded the French people to salvage their household items—clocks and bells—for munitions.

Monge served as Revolutionary Minister of the Navy, an impressive title and a real achievement for someone of such humble origins, even though the French navy was a relatively impotent force. In that post, he first met the young officer Napoleon Bonaparte and bestowed on the still-unknown soldier some small, but apparently memorable bureaucratic favor. Years later, Napoleon, now a general, invited Monge to join him in Italy and charged him with overseeing the removal of loot from Rome and the Vatican. Thereafter the general and the geometer became lifelong friends. "Monge loved me as one loves a mistress," Napoleon said later. The young general was so enamored of his older, pugnacious friend that he routinely ordered military bands to strike up the revolutionary national anthem, the *Marseillaise*, before meals they shared, explaining "Monge is an enthusiast for it!" Monge would always dutifully bellow the words to the national song before eating.

Devoted as he was to Napoleon, Monge was at first reluctant to go to Egypt, fearing it would wreck his marriage. When Napoleon's summons reached him in Italy, he offered his age as an excuse, writing back that "I am not very young. Leave me among the mortals to admire your talents, appreciate your services, and sing your glory."

Bonaparte wouldn't take no for an answer, Olympian flattery notwithstanding. He wrote back to say that if Monge didn't come right away, and with the Vatican's Arabic-character printing press in hand, he would send a squadron up the Tiber to get him. When she heard he was considering another expedition, Monge's wife, back in Paris, was furious.

"Old fool," she fumed, "Running around the world at your age? Haven't you a good wife, daughters who love you tenderly, a sufficient fortune? You want to get killed! Instead of leaving for you don't know where, couldn't you dream of finishing the rest of your life peacefully, enjoying the things you possess?"

Monge showed this letter to his friends in tears, then wrote his wife back, explaining that he simply had to follow Bonaparte whither he would lead.

After five days marching in the burning sand, and countless deaths, the French did finally see a fantastic sight on the horizon that was not an optical trick: thousands of men, on horseback. From afar they resembled calligraphy on the horizon, dark silhouettes perched easily on their cantering Arabians, trailing gold and scarlet silk. Up close, they proved to be huge men—they had, in fact, been specially selected as boys by experts who examined their limbs, eyes, and teeth, searching for pedigree, looking for a specific *breed*, found chiefly in the Eurasian steppe between Central Asia and the Adriatic Sea. A few were black, stolen from their mothers in Africa; most were white, with blue eyes and great blond beards. They were, wrote one Frenchman, "beautiful men, with a complexion of lilies and roses."

In Cairo in 1798 the trade in human flesh was as ordinary as the trade in spices and horses. The city served as a marketplace for animals, raw materials, and human beings from deeper in Africa. But black slaves, shipped up the Nile in chains or marched across the desert in caravans, and sold for one rifle if female, two rifles if male, were different from the Mamelukes, who were slaves by pedigree only.

Mameluke society was so androcentric that its women routinely practiced abortion and sometimes even dressed like teenage boys to stay attractive to their men. They replenished their ranks not by biological reproduction, but with boys aged eight to ten years old whom they bought in the Caucasus and trained as warriors. "Strangers to each other, not bound by those natural ties which unite the rest of mankind, without parents, without children, the past has done nothing for them, and they do nothing for the future," Volney observed. "They become ferocious from the frequent murders they commit."

Volney exaggerated and romanticized the Mamelukes, but it was hard for any European writer to resist that urge. Their look and legend fascinated Europeans for centuries. Long after they were gone, artists depicted the turbaned Mameluke on horseback, silks wafting on the wind, scimitar aloft, as a symbol of Arabia. Gorgeously dressed, impervious to the withering heat, each man was an equestrian treasure chest. The Mamelukes adhered to the nomadic tradition of transporting as much of their worldly wealth as they could carry in gold, silver, and jewels on their bodies, or hidden beneath their saddles. For ceremonial occasions, they decked themselves in layers of linen and silk shirts, gowns from neck to ankles, and outer cloaks of scarlet satin lined with beaver, embroidered and belted with gold. Beneath all that, they wore giant pantaloons of bright Venetian cloth, the fabric on each leg wide enough to swath an entire body.

The Mamelukes were also armed and dangerous. They were trained from childhood to sever heads with one swipe of a sword while galloping with reins in their teeth. In battle, they carried English carbines, which they discharged first, then slid under a leg. Next they fired their pistols, flinging them over their shoulders to be recovered and reloaded by attendants following on foot. Then they pitched *djeriads*—sharpened javelins made of palm branches. Finally they engaged the enemy with the jeweled scimitar's thirty-inch blade, all while galloping at top speed.

The Mamelukes revered nomadic life yet felt a strong sentimental tie to the city of Cairo, all the while loathing and fleecing its native inhabitants.

When the French arrived, Egypt officially belonged to the Ottoman Empire, but the Mamelukes controlled, and profited by, the country's day-to-day administration. In return for the heavy taxes they imposed on the population, Mamelukes were expected to protect Egypt from foreign invasion, and they had successfully repelled the Crusaders and then the Mongols in the thirteenth century. Because of this track record, the great Muslim scholar, Ibn Khaldun, called the Mamelukes "the saviors of Islam." From 1260 to 1517 they reigned over Egypt, as part of the Baghdad Ayyubid and Abbasid dynasties.

The Turks defeated the Mamelukes in Egypt in 1517, in part because the Mamelukes held to a code of honor that forbade them from using firearms offensively. (That part of their code was abandoned by the time the French confronted them.) They also eschewed modern weaponry in favor of old-fashioned equestrian warfare with swords. Superior Ottoman cannon tore them to pieces. After that battle, the Turks entered Cairo and hunted down and beheaded thousands of Mamelukes, decorating the streets with impaled Mameluke heads. Within a year, however, the Ottoman sultan came to terms with the Cairo Mamelukes, putting them on the payroll and allowing them to administer the country. Thus did they continue as the de facto rulers of Egypt for another 272 years.

When the French arrived, the Turks nominally ran the country, but the Sublime Porte was a long way from Egypt, and the Mamelukes were living off the fat of Cairo's tax base, and employing 90 percent of the populace as peasants on their lands. They had divided Egypt into twenty-four provinces ruled by twenty-four beys (chieftains) and had developed a complex society, separate from that of the Arabs they ruled. They employed Coptic Christians as tax collectors and scribes, creating an additional buffer between themselves and the people over whom they ruled.

By 1798, turf wars between the beys and uprisings by the native Egyptians, not to mention the Mamelukes's own traditional refusal to propagate themselves other than through purchasing boys, had so weakened them that the Mamelukes were in serious decline. With Russia in control of the Caucasus, it was becoming harder for them to buy fresh boys from the region. Recruiters had turned to older men from other areas, including Christian Albanians and Greeks, to replenish their ranks. Historians debate the number of Mamelukes living in Egypt when Napoleon arrived. When he confronted them in the most decisive battle, he faced no more than 12,000 elite horsemen, probably fewer.

Mameluke beys still controlled Cairo in 1798, building fabulous palaces while the rest of Egypt starved. Two beys uneasily shared top control. Ibrahim Bey, whose honorific title was *sheikh el-beled*, (elder of the town) was the elder in years of the two and, compared to his fellow ruler, was rather dry and scholastic. Murad Bey, *sheik el hajj* (commander of the pilgrimage), was the younger and more flamboyant of the last two beys. Murad lived in a palace near the pyramids at Giza, where he indulged his taste for literature, music, and chess. The Egyptian writer Abd al-Rahman al Jabarti, who chronicled the French occupation from the Arab point of view, described Murad as a complex epicure—a man who preferred associates who were "hard, brave, and cruel" but who also chose men of letters and taste as his confidants. He enjoyed the good life, and spent long periods without leaving the pleasures of his Giza palace. He was famously generous to his Mameluke followers, but fleeced the native Egyptians, as had his predecessors.

The French artist André Dutertre, who sketched most of Napoleon's scholars in profile during their time in Egypt, sketched Murad's portrait at his home, capturing his piercing look, bushy pale beard, flowing robes. In the picture, printed in the scientists' final book on Egypt, Murad is half-reclining on a bank of pillows, holding a horsehair fan or flyswatter, his scimitar nearby, and the plains of Giza visible through the arched window behind him. Dutertre's likeness did not include the long, pale scar on Murad's cheek, badge of battles past.

The French had legitimate reason to believe Egypt would be easy to take. European travelers and merchants in Cairo, while they feared the depredations of the Mameluke tax collectors and the routine abuses and brutality of the social system, had been assuring Europeans that Egypt was essentially undefended and ripe for the taking since the time of Louis XIV, years before the French Revolution.

In his war planning, Napoleon dismissed the problem of nominal Turkish rule of the country. He believed he could get around that by persuading the Turks he was eradicating the Mamelukes for their benefit. To make that argument, the French government was supposed to send a foreign minister to Turkey with assurances that the French were not really at war with the sultan—this while the still code-named

"Army of England" sailed toward Alexandria. The minister, though, never went to the Sublime Porte, but stayed home in France. When the Turks learned of the French war plans in Egypt, they promptly jailed the French consul in Constantinople and declared war on France. Napoleon's failure to predict the strength of the Turks' attachment to Egypt was one of the worst miscalculations of his career.

Mameluke brutality did offer Napoleon an excuse for invasion tailored to the Egyptian people. In his first Alexandrine proclamation, printed in Arabic, and distributed to the religious scholars (among the few Egyptians who could read), Napoleon assured the Arabs that the French were going to free them from the Mamelukes. Besides ordering his officers to consult French translations of the Koran to find common ground with the Arabs, Napoleon distributed Arabic translations of Tom Paine's *The Rights of Man* to the learned class of Egyptians, to further bolster what he hoped would be the Arabs' latent revolutionary spirit.

Napoleon's atheist soldiers encountered the Mamelukes just after they reached a copious water supply in the form of a sluggish branch of the Nile at Rahmanieh, west of Cairo. The sun-addled French plunged in fully clothed, some drowning immediately due to the weight of their arms. They guzzled the leech-infested water (the hideous details of this ingestion are the subject of an official medical report in the scientific corps' records), and wrestled canoe-size crocodiles. They crawled out and gorged on watermelons, the only food in sight, which soon produced their first bouts of dysentery. Here, at the river's edge, men who still had their strength made up a mocking song that summed up the mood of the rank and file, as well as of the officers.

> *L'eau du Nile n'est pas du champagne!*
> *Pourqoui vouloir faire campagne*
> *Dans un pays sans cabaret?*

From this watermelon field the foe was finally spied. At first he was just another mirage on the horizon, a trick on the eye. Then drums boomed. Trumpets blared. Chants rang out. Twelve thousand Mamelukes—the entire corps of the fighting elite—accompanied by thousands of footmen and aides, prepared to meet the infidel. In unison, they invoked the name of Allah and focused on the pleasures of eternal life promised to the faithful who fall in battle.

With the Pyramids silhouetted in the distance, Bonaparte shouted, "Soldiers, from the summit of yonder Pyramids forty centuries behold you!" How he was able to so nearly guess the real age of the Pyramids a half-century before archaeologists determined it is one of the legendary Napoleonic mysteries. The French soldiers formed tight, elongated squares and waited for the Mameluke charge.

The ensuing clash was short and bloody. The Mamelukes had never seen such a battlefield maneuver, and between the squares and the superior French weapons, they were particularly ineffective. Murad Bey had ordered his men to charge and encircle the French. The French held their fire until the Mamelukes were less than fifty paces away, then let loose, killing the foe in waves. Sometimes the fighting was so close that the flash from the French musket barrels set fire to the Mamelukes' silk robes, who fell and burned to death just yards from the impregnable French squares.

It took an hour for Murad Bey to realize the futility of his horsemen's efforts and order a retreat. In a last desperate attempt to save themselves, the Mamelukes started throwing gold coins from their purses on the ground before the French, hoping to distract them. The tactic failed.

Several thousand Mameluke corpses littered the field, and the French fired at survivors who tried to flee across the Nile. At least a thousand drowned in the river and another six hundred were shot in the water. The French troops dove in after them, fishing out Mameluke corpses to get their hands on the fabled treasure.

By four-thirty in the afternoon, the battle was over, but, as the

French would soon learn, not the war. Murad Bey fled south toward Upper Egypt with some three thousand Mamelukes (the other survivors gave up or broke ranks), leading one French general on a nine-month wild goose chase to the edge of the Sudan and back.

Very few French soldiers died in the so-called Battle of the Pyramids. Monge and Berthollet, however, nearly perished in a small skirmish on the Nile. Before the battle, Napoleon had tucked his favorite *savants* out of harm's way on one of a small number of French boats in the Nile, but a Mameluke flotilla, manned by Greek sailors, attacked. Warriors began boarding a French boat with scimitars aloft and soon Mamelukes were holding up severed French heads by the hair for the benefit of the infidels on other boats. Watching the spectacle and fearing the same hideous fate, Berthollet filled his pockets with stones, preparing to quietly drown himself. Monge jumped into the action, reloading guns and shouting encouragement to the French troops. In the nick of time, Napoleon heard of the fighting on the Nile and rode with some guards to the rescue. For years, Napoleon blamed his failure to annihilate the Mamelukes completely on the need to save Monge and Berthollet that day.

Napoleon's easy victory over the Mamelukes encouraged his spirit and was good for the troops' morale. As they marched unimpeded through Cairo's gates, to the ululating shrieks of terrified women, the French were entering a fool's paradise. The archaic, routed Mameluke equestrian force would be the easiest foe they would face in Egypt. From that day forth, their enemies would be national entities more their own size—the British and the Turks—who, joining forces, weren't about to let the French take Egypt easily.

The British-Turk alliance, though, was still on the far horizon, and not even a flicker of a possibility in the general's mind. Celebration was in order. That night, Bourrienne, Napoleon's secretary, sent a letter to Napoleon's brother, still near the coast. "The General enters Cairo this evening. Send our baggage and the wine."

Despite his painless victory, Napoleon admired the enemy's grit. "Ten thousand Mamelukes could easily fight and win against 50,000 Turks," he wrote later. "I could not imagine what I could do using a fistful of these warriors." He didn't get a fistful of them right away, but before the French left Egypt, the Mamelukes and the French would ally against common enemies, the Turks and English, and a decade later in Europe Napoleon had a Mameluke division fighting for him in Spain. The general did soon *own* two Mamelukes himself, though. When Cairo surrendered to the French a few days later, the leading Arab sheik presented the conquering general with the customary tribute: a magnificent Arabian steed, and two Mameluke slave boys, Roustan and Ibrahim. Slavery was illegal in France, but the general graciously accepted these human gifts anyway. Ibrahim eventually worked for Joséphine in France under the name Ali, and Roustan would become Napoleon's personal bodyguard and Oriental prop in Europe. This Mameluke— one of the last of the breed—lived out his days in France, attending every procession, journey, and battle beside the French leader in the coming decade, kitted out in fabulous costume, a living memento of the Egyptian expedition.

CHAPTER 3

THE INVENTOR

He had all the Arts in his hands and all the Sciences in his head.

—Gaspard Monge,
eulogizing Nicolas-Jacques Conté

Alexandria and Rosetta, July 1798

The scientists floated in Abukir Bay for two days without any news from shore apart from the pock of occasional gunfire and what they could glean from peering at the beach through spyglasses. They saw the silhouette of a poor, dusty town of six thousand souls and, they would soon learn, an almost equal number of feral dogs. Ramshackle mud and brick hovels lined the narrow streets. Other than Pompey's Pillar, rising indignantly elegant on a garbage-strewn hill in the distance, the scientists couldn't detect a single intact remnant of the classical past, certainly no stone or beam from the fabled library of Alexandria, repository of ancient learning. Peering at the shore, Denon noted: "It was not variegated with a single tree, or a single house. He [Denon always referred to himself in the third person] thought it wore the appearance, not of the melancholy of nature, but of her ruin, of silence and of death." The scholars all felt their classical expectations dashed. "We were looking for the city of the Ptolemies, the library, the seat of human knowledge," wrote the architect Charles Norry. "And we found instead ruins, barbarism, poverty, and degradation."

Taking his turn squinting at the crumbling silhouette of the city, the student engineer Édouard Devilliers had a crisis of confidence. He began to wonder exactly what service a student schooled in the art of surveying could possibly perform in this bleak place. His elders reassured him that his skills would indeed be needed.

On the fourth of July, cannon shot finally echoed across the water from shore, a signal to the scientists to disembark. Unceremoniously deposited on the beach with their luggage a few hours later, the scientists blinked in the inescapable noon sun and looked around. Heaps of garbage and sand encircled the old city walls. The town seemed deserted, and no human sounds broke the eerie quiet of midday. The houses were all shut up, the inhabitants barricaded within, but peering out through the tiny holes in the *mashrabiyah* (latticework) windows, watching the invaders. Some scholars feared the Alexandrines might shoot at them from their windows. They soon learned they had more to fear from sun, dogs, and insects.

Looking around for a welcoming committee and finding none, roasting in their hot clothes, the men dragged their luggage through the sand and clambered over the city walls into a ruined cemetery. "The first scene that presented itself was a vast burying ground, covered with innumerable tombs, raised in white marble upon white earth," Denon wrote. "A few meager women, wrapped in long and ragged garments, looked like so many ghosts wandering among the graves."

The first afternoon in Alexandria, the *savants* fanned out and searched for shelter from the sun. Some collected palm fronds to make temporary lean-tos, as the soldiers had done on the beach. Others found their way to the European quarter of town, and persuaded the tiny community of merchants to give them lodging, packed ten to a room on the floor. When they found shade, they lay down to rest on their baggage, with ample time to contemplate the disparity between their idea of Egypt and the reality.

Denon didn't try to sleep. As the sun set, he had to walk around, feel terra firma at last beneath his shoes, get his bearings. Throughout the first night, he stalked the deserted streets. The silence was uncanny. As he wandered, Denon found the city "a melancholy one," with "nothing that resembled Europe and its cheerfulness, except the noise and activity of its sparrows." Feral dogs eventually chased him through the streets and up a wall, where he crouched until daybreak and the arrival of some French soldiers. Sitting a few feet above snarling jaws for hours, the artist contemplated the Alexandrine canine. He later recorded his impressions. "In its dogs, he could not recognize the friend of man, the faithful and generous companion; they were here morose egotists, strangers to the master under whose roof they were sheltered; they mistrusted him whose abodes they nevertheless defended, but whose dead body they would without horror devour."

In the morning, with the sea breaking in towers of spume on one side, and sand as far as they could see beyond the town walls in every other direction, the civilians shed their coats, shaded their eyes, and got to work. Their first order of business was locating potable water. The cisterns, black and stinking, contained liquid of dubious quality. Dry-mouthed and dizzy with thirst, they gulped it down anyway. Good water would always be scarce in Egypt, and the engineers were almost immediately set to work seeking wells, or digging ancient wells deeper in search of water beneath layers of putrefaction.

After water, food preoccupied them. The army hadn't bothered to provide the scientists with any rations—the soldiers were barely fed themselves. In Europe, Napoleon's army ate by forcing people it "liberated" to pay for their own liberation with food and wine. In Italy, where butchers, bakers, and vintners abounded, this worked nicely enough. In poverty-stricken desert Egypt, the tactic would fail miserably. On the best days, the luckier men ate bits of camel meat and rice with their ration biscuits; more often, especially when they were away from the cities, they subsisted on dry biscuit and green dates.

On his first day in Alexandria, Devilliers met a soldier in the street and offered to buy some of his rationed biscuit. The soldier, appalled to find a fellow Frenchman begging from him, insisted on not being paid. This act of generosity from soldier to *savant* would become rare as rain in the months to come.

The desert heat stupefied the new arrivals, although they shed their wool coats and eventually acclimated; some even grew to love the climate. The insects were another matter. The civilians had never met such abundant flying, biting creatures—a phenomenon that thrilled the naturalists but no one else. Mosquitoes, fleas, and tiny gnats attacked at all hours, and the flies, which swarmed into all cavities—nasal, oral, ocular, aural—were impervious to swats. Invisible bugs were more viscerally annoying. Diarrhea and dysentery soon afflicted everyone. Simply opening one's eyes could be a hazard. Dry heat, sand (they had arrived at the season of the *samoon*, unpredictable, sudden, and powerful sandstorms), and bacteria caused a painful eye infection they called ophthalmia, which attacked nearly every man at one time or another. Sufferers said it felt like needles in the eyeball. It left men's eyes running with pus and temporarily blinded them, sometimes for weeks.

The disoriented students complained copiously about the privations in the first days, but the elder scholars accepted the difficulties of Alexandria with equanimity. To the older men, the hardship of daily life paled against a sense of discovery. From their first hours on land, the elder scientists went about collecting, measuring, and exploring, while the addled students still couldn't imagine how they would ever be fed or housed. "They were already busying themselves with works; our living had not yet been provided for, and we were most absolutely abandoned," was how engineering student Prosper Jollois saw it. "What a beautiful subject of a caricature," Jollois complained, "to see the *savants* at the Column of Pompey, painfully measuring the dimensions, and when returning, searching with avidity cisterns where they could quench their thirst, while waiting to satisfy their hunger!"

British caricatures of the French invasion did indeed use Pompey's Pillar and the scientists measuring it, as the visual backdrop to satirize Napoleon's expedition in Egypt. A massive polished red granite column, thirty yards in circumference, Pompey's Pillar towered nearly a hundred feet in the air. The pillar actually has nothing at all to do with Pompey—Caesar's vanquished foe. The soaring classical column is one of dozens that once formed the Serapeum, or Temple to the sun god Serapis, erected by some of the last of the pagan Greek rulers of Egypt in 300 B.C. The temple had been a gleaming, fabulous place, with an enormous elevated sanctuary decorated with plates of gold, bronze, and silver. A deep-blue, luminous statue of Serapis, fashioned out of filings of gold and ground fragments of topaz, sapphire, and emerald, served as the centerpiece.

Christians attacked the pagan temple in A.D. 391, and the Romans later took a pillar from the wreckage and placed it upright to honor a Roman emperor, Diocletian. Medieval Europeans erroneously believed that Pompey, defeated by Julius Caesar in 48 B.C. was buried under the pillar and that his head was entombed at the top, inside the capital.

The *savants* knew nothing of the true provenance of the pillar. As with most everything they encountered in Egypt, they attempted to understand it using the means of inquiry they relied on as men of science, including precise measurement and careful analysis of substance. Just days after arriving, a large group of scholars tramped off to the pillar at five A.M., before the heat of the day, accompanied by a military escort. Using a kite to raise a pulley over the top, they then hauled a small sailor up one side, who installed a stronger rope capable of supporting heavier men. They ascended individually, one after another, inspecting the slippery rock on the way up. Geologist Dolomieu remained on the ground, analyzing the building materials. He later correctly pronounced the rock at the base to be more ancient than the shaft. In fact, the pillar was standing on an even more ancient Egyptian site.

In the same methodical way, the scholars inspected other antiquities. On their first full day in Alexandria, Monge and Geoffroy Saint-Hilaire examined the two obelisks near the pillar (one of them, now known as Cleopatra's Needle, resides in New York's Central Park; the other stands on the Thames Embankment in London). Denon and Dolomieu discovered a huge, hieroglyph-covered sarcophagus inside a mosque that had previously been a Coptic church. The French—and later the British—suspected that the sarcophagus once held the remains of Alexander himself. In fact, it had housed the bones of an Egyptian king named Nectanebo II. By 1798, it was being used as an ablution tank by the Muslim faithful. Inside the same mosque, another *savant*, the opera singer turned expedition musicologist René Villoteau, unearthed sheet music for medieval chants dating to 825 and attributed to Saint John Damascene, the founder of modern Greek music.

Soon, *savants* with measuring instruments and notebooks fanned out across the city. Alexandrines slowly emerged from their houses, now more confused than terrified. Who were these unarmed men in frock coats and spectacles, carrying strange tools made of wood and glass, pencils, and paper?

The scholars had been warned that desert Bedouins who wandered in and out of the city could be dangerous, but by day the scientists found them more likely to be friendly merchants who sold camels and food. At night, however, the Bedouins picked off solitary stragglers. One of the scholars, an Orientalist and interpreter named Panhusen, disappeared in the streets of Alexandria the day after the savants landed and was never seen again, presumed murdered.

A week after arriving at Alexandria, a group of scholars were so comfortable in Egypt that they didn't bother to wait for military accompaniment before setting out to explore the desert beyond the city walls. The tailor Bernoyer watched in amazement as a group of thirty Bedouins on horseback escorted some six wayward *savants* back into Alexandria from the desert, unharmed. "Only the Arabs' humanity saved them," Bernoyer wrote. "Bonaparte was pleased at their intrepidity and the Arabs' kindness."

This friendly exchange between scholars and natives was more typical of the civilians' interaction with the Egyptian people than the army's, which tended to involve exchanges of bullets and swords. Where the military saw enemies in the Egyptian faces, the scholars regarded them as worthy of study. They took notes and made sketches on physiognomies, dress, and houses and eventually tried to document social and sexual mores.

In Alexandria, they first grew accustomed to the sound of the mournfully intoned calls to prayer, issuing night and day from the minarets, pulsating in rhythm with the dry heat radiating off the sand. On their rounds of the town, the scholars stopped at doorways and peered in, wide-eyed, at children chanting and swaying in the Koranic schools, reciting verses. Atheists, they scorned the rote religious education system—"this bizarre spectacle," as one of them put it.

For the first time in modern history, a large group of Westerners observed Islam as a lifestyle. The French scientists were confused by the culture's asceticism, and ignorant of some of its basic tenets. The apparent inflexibility of the religion, the poverty of the masses, and the treatment of women revolted them. Napoleon had handed out Korans and instructed his soldiers to be tolerant of Islam, and when it suited them, they were. Other times, they quartered horses in mosques and disrespected what the people most venerated. The scholars, at least, regarded the holy places and objects as worthy of study, and some learned Arabic well before they left.

One scientist who had to learn the Egyptian language in order to do his work was the inventor and chemist Nicolas-Jacques Conté. Conté worked side by side with Arabs in his makeshift factories, studying their fabrication methods and teaching them what he knew. Throughout his life, Conté invented useful things and imagined other machines and devices that he himself couldn't fabricate but which would eventually help move the world into modernity. A farmer's son from the village of Saint-Cénary in the Savoie region, Conté was a boy of nine when he carved a working violin. At the age of twelve he replaced a sick church-painter in his town church and made ceiling panels of the saints with such vividness and expression that experts from around France came to gape. Conté was not, however, destined to decorate. The boy thought maybe he wanted to be a gardener.

In the lean years after the Revolution, men with inventive talents were badly needed to fill in the gaps that siege, war, and broken supply lines created. In Paris, in his twenties, Conté worked as an apprentice to the famous painter Greuze. As the Revolution approached, Conté worked nights in a chemistry lab and soon stopped painting altogether and started making things—tools, paints, gases—that the kingless country would need. Before long, the revolutionary government identified him as the man to go to when there was a need that no one else knew how to fill.

The stories of Conté's indispensability in revolutionary Paris are many. Some have the apocryphal flavor of heroic legend, but many are probably quite true. During the darkest days after the Terror, with France at war all over Europe, the revolutionary leaders wanted to cast commemorative medals. Looters had smashed the old royalist machinery and raw-material shortages prevented the creation of new machines. Still, medals were needed, to honor the brave revolutionaries. A committee called on Conté. He examined their efforts and said he was not surprised that they had failed, given that they were missing a small tool. "Which tool?" he was asked. He replied that he did not know, but that one was missing. The committee asked him to invent it, and he went home, shut himself up in his laboratory, imagined an instrument, fabricated it that night, and brought it to the committee the next morning. On the first try, it cast a perfect medal. From then on, Conté added to his duties a series of studies on machinery used to make coins.

Conté's best-known contribution to France had to do with a small, mundane object crucial to day-to-day life. In the 1790s, France, at war with England, the major source of graphite, had run out of pencils. Conté invented a new compound from which to mass-produce pencils by kiln-firing a mixture of clay and graphite. The pencils sealed his fame. The world still uses a version of Conté's hybrid pencil today. Among his other inventions: an engraving machine that could replace the work of eight men, and a new kind of paint whose colors were so durable they would stay bright for centuries.

Like many inventors before and after him, Conté was fascinated with the idea of manned flight—in his era, hot-air balloons. Conté lost an eye trying to find a cheaper kind of gas to keep French military balloons aloft. He mixed and tested balloon gases in his own lab; working alone one night, experimenting with hydrogen, a sudden gust of air blew the gas toward a lamp in a corner of the room. After the explosion, when Conté regained consciousness, his face was embedded with glass fragments and he was blinded in one eye. For the rest of his life he wore a black eye patch, but he was rewarded with the position of chief *aérostier.*

A patriot who revered Napoleon and was devoted to the army, Conté was primarily an inventor, who foresaw uses for balloons beyond the needs of the military, among them weather prediction. He brought a few balloons with him to Egypt, but only sent them aloft as a public spectacle in an unsuccessful attempt to win over Egyptian hearts and minds.

Conté was an optimistic, humble, and innately cheerful man who invented solely for the pleasure of making ideas manifest in useful things. He only reluctantly accepted patents and refused to take payment. He was eventually asked to create France's first school of industrial and applied arts. He was overseeing this new institution when the Directory called him to accompany Napoleon to Egypt as a member of the *corps de savants*. His directive was to help the army and to study the factories of Egypt and improve them wherever possible.

Soon after they arrived in Egypt, the scholars divided into two groups. The first, mostly engineers and architects, settled in Alexandria, and Conté was enlisted immediately to help supply the army's various daily needs. Artists sketched the coastline around Abukir Bay for tactical planning purposes. Surveyors mapped the city, engineers searched for water, astronomers determined latitudes and longitudes to help with mapping. Doctors tried to set up hospitals. The expedition's lone musicologist was called into service to play music for one of the generals.

The mappers had the most dangerous duty, owing to their extended hours spent measuring isolated areas. In two months they finished the most detailed cartography ever made of Alexandria at the time, though some died in the process. Two Frenchmen were killed when Arab insurgents attacked a group of geographers doing survey work near the gate at Pompey's Column.

The soldiers watched the intellectuals on their rounds, simultaneously bemused and annoyed. One captain wrote to his wife that although Alexandria was abominable, the scientists and artists appeared happy and fascinated. "That is to say, our privations don't seem to affect them, the ruins of antiquity nourish them enough." Relations between soldier and *savant* had been strained on the ships; on land they deteriorated further. The *savants* needed the soldiers—they couldn't explore very far afield without armed guards, even with friendly Bedouins leading them home—but the soldiers had little need for their services, besides what the mappers and water-finding engineers could do. Napoleon, though, would soon be finding ways to make all of them useful. He had already ordered some artists to start engraving the names of the recently dead French onto the column in Alexandria.

The list was growing faster than they could carve.

A larger group of *savants*, including all of the naturalists, students, and artists, soon moved up the coast to Rosetta, a delta town with more water, better food, and comparatively luxurious conditions. On July 8, as the army marched off into the desert, the first group of *savants* headed for Rosetta. Their departure was chaotic, with every man fending for himself. Scholars lost their baggage in the transit, which involved several changes of boats, walking, and riding. The students were temporarily left behind when their professor, Fourier, snatched a berth on a military transport for himself, leaving his charges to make their own way. Geoffroy Saint-Hilaire took pity on the younger men and arranged passage for the students on a shabby merchant vessel.

The scholars were much happier at Rosetta, where, Geoffroy Saint-Hilaire wrote, they found a town as gentle and welcoming as Alexandria was brutal and unfriendly. "It is an earthly paradise!" he exclaimed. "The temperature of our May reigns here, and it is raining!" Besides the miracle of rain, the scholars easily found clean water to drink. They slept in fine houses recently abandoned by wealthy Egyptians fleeing the French invaders. Geoffroy collected animals within the city limits—mostly pet birds and reptiles. He shared a luxurious house with twenty members of the Commission. Three Maltese slaves waited on them—the French had apparently not liberated *everyone* in their path, contrary to propaganda. Here at last were the scented gardens and shaded courtyards Savary had promised.

The revolutionary month of *messidor* (harvest) passed into *thermidor* (heat). The French had been in Egypt for a month, and the scholars settled into a sort of routine on the coast, waiting for the all-clear from Napoleon to proceed to Cairo. In the bay, the French ships and sailors were also waiting for word from the general. The entire fleet was anchored a mile and a half from shore, thirteen battleships forming a mile-long line, linked together stern to bow by cables that stretched fifty yards. This linked line of vessels, containing everything from the French arsenal to soon-to-be badly needed medical supplies, was intended to keep British ships from slipping to the land-side of the French until Napoleon sent word to unload.

Napoleon's secretary, Bourrienne, wrote later that Napoleon had planned to unload the ships as soon as Egypt was secured and send them back to Toulon where they could rejoin the rest of the French fleet and attack England across the Channel. But the General never got around to ordering the French ships into some more protected cove before leaving for Cairo.

After several uneventful weeks bobbing in the bay, the French admiral in charge, François-Paul Brueys d'Aigailliers, felt comfortable sending more than four thousand of his sailors to shore to bring back wheat and other desperately needed supplies. Brueys would die regretting anchoring his ships so far from land in an unprotected bay, and his lack of manpower, when the guns started firing.

Going about their business on the steamy afternoon of the first of August on the conventional calendar, most of the scientists on land didn't notice the British fleet creeping up on the horizon, because the line of French ships blocked their view. They didn't need spyglasses, though, to see the plumes of black smoke filling the sky when the historic Battle of Abukir Bay commenced. One-third of the fleet's sailors could only watch in horror from shore as Admiral Brueys and his linked French ships were caught off guard. With ease, the British fleet snuck around the French line and began blasting away, as the superior British navy outmaneuvered the clumsier French sailors.

The French fleet was wrecked in one night, famously capped by the explosion of Napoleon's floating salon and arsenal, L'Orient. That blast shook the ground for miles on land, fire lit the night sky briefly, and then, as water swallowed the remains, silence and darkness returned. At sea, even the British observed a spontaneous moment of silence in awe at the force of the explosion. The ship sank to the bottom of the sea in bits and pieces, carrying hundreds of men, along with Napoleon's royal bed chambers, the gold, silver, and jeweled saints' heads snatched from Malta, and whatever was left of the expedition's library.

One thousand seven hundred men died at Abukir Bay. The British took more than three thousand French prisoners, most of them wounded.

Without communications, the scientists in Alexandria and Rosetta did not realize the full enormity of the disaster until weeks later. For the first few days they were able to hope the French had prevailed. Many scholars witnessed the deafening explosion of what turned out to be *L'Orient* from a tower at Rosetta, but no one knew whose ship had sunk until much later. In his journal, Prosper Jollois recorded the sight, without knowing it was Napoleon's ship he watched explode. "The night was dark. We caught sight of numerous flashes formed by the light of the cannon. A horrible carnage continued. Oh! But it is terrible the thought of a naval battle! I was absorbed in these painful reflections when a continuous white glimmer, that grew by degrees, made an impression on our sight. It grew rather quickly, and soon, we were no longer in doubt that it was a ship on fire. It did not stop firing its broadsides. Finally, the fire having probably reached the powder-magazine, the ship exploded. Nothing more appalling and more beautiful!"

On the evening of the second day after the battle, everyone celebrated on the weight of a rumor that the French had prevailed, but a day later the horrible truth was confirmed. "Apparently the French have no more fleet!" Devilliers wrote. "I was told that everything that was said about the English the day before, was in fact true of the French fleet—they lost!"

Ten days later, the scholars at Rosetta reckoned with the true horror of the disaster they'd witnessed from afar when the flotsam and jetsam of the battle washed ashore all at once. In a single horrifying day, thousands of rotting human carcasses littered four leagues—approximately sixteen miles—of sand. Some bodies were still intact and in uniform, others stripped of every thread, pathetically naked, nibbled by fish down to scoured bone. "Our beasts of burden shuddered to walk among them. At the least blow, the bodies released the odor of rotting flesh," wrote one military observer.

"The odor that reigns all along the coast, for three or four leagues is foul," Jollois wrote in an entry dated August 13. "One sees piles of six or seven cadavers, sometimes a leg, sometimes an arm sticking out of the sand; it is truly a horrible spectacle."

The *savants* watched as the Arabs began setting small fires all along the coast, burning the debris of the vessels as it washed ashore. The Arabs were not busy making funeral pyres for the French dead. They were burning the wrecked wood to salvage nails. Egyptian peasants had little access to metal, let alone nails. Most of the raw and refined materials taken for granted in France were scanty in Egypt. As the scholars surveyed the ghastly beach scene, and tiny fires burned the last scraps of their lifeline to home, they understood that there was little difference now between them and the native inhabitants of Egypt. From that day forth, the French scientists were going to be scavengers too, in this land of sand and ruins, flies and plague, and when their tools broke or wore out, they would replace them with what they could fashion from papyrus, palms, scrap metal, and reeds.

The scholars knew that everything they thought themselves entitled to as Frenchmen—weapons, food, gear, clothing, books—lay at the bottom of the sea. Each passing day, they remembered more necessities of life that they would not have: medicine, buttons, surgical instruments. Shoes. Glass. Pencils. Worst of all, their means of receiving and sending letters was lost as well. "From then on we realized that all our communications with Europe were broken," wrote the engineer

Étienne Malus in his memoir. "We began to lose hope of ever seeing our native land again."

Having no choice, the scholars returned to work. They collected birds, insects, and plants, measured colossal statues, dissected mummies, and denied what lay ahead: Paris would never receive their reports. Their desperate requests for more supplies would wind up in British hands or sink to the bottom of the sea. They had no way home. They were trapped in Egypt with nothing but the tools in their hands and the shirts on their backs. When those shirts and tools wore out, they relied on one man to keep them supplied.

Nicolas Conté was forty-four years old and the director of a new school for applied industrial arts when he was called to join the expedition to Egypt. Besides his post as director, he left his wife and small daughter behind. Even before the disaster at Abukir Bay, Conté had been proving useful to the army. In Alexandria, for example, the general in charge noticed that despite the dry air, his guns were rusting from morning dew. Conté quickly arranged to have a few weapons bronzed, which proved protective. The army bronzed the rest with material brought with them from France.

Berthollet called Conté "the column of the expedition and the soul of the colony." At Cairo, Conté eventually installed and oversaw a small village of workshops where he fabricated everything from gunpowder to cloth, paint, and glass instruments. Using only indigenous materials, he created all sorts of machines, including a printing press, a coin press, geometry machines, engineering tools, and even trumpets for the army. He built a smelting foundry to make steel and then produced sabers. Among his greatest contributions to the military were devising a way to bake large amounts of bread, designing windmills to grind wheat, and creating a factory to make a light woolen cloth used to make new uniforms when the old ones turned threadbare.

Conté carried out his military duties without neglecting his own painting, experiments, and fieldwork. In France he'd invented a new type of barometer, which he brought with him to Egypt, where he used it to measure slight differences in atmospheric pressure. at different elevations. By means of this device, he was able to measure the height of objects, including the large pyramid at Giza, which he calculated to be 428 feet, the same figure later deduced by more sophisticated instruments.

Because Conté employed large numbers of Arabs in his workshops, he interacted more intimately with Egyptians than most of the French. Troubled by his inability to communicate, he taught himself Arabic. He visited and studied local Egyptian workshops and carefully noted their means of smithing copper, tanning leather, weaving, glass- and pottery-making, tool-making, and Egyptian uses of various native natural resources. He produced fifty watercolors of these workshops and their inhabitants, as well as numerous individual portraits of workingmen and women. These images, published in the final book, are among the most evocative depictions of Egypt in 1800 available. In his biography of Conté, the engineer and expeditioneer Edme-François Jomard described how carefully the one-eyed inventor surveyed his subject before putting brush to paper. "He had to render each scene carefully—and it was not easy—so as to show each mechanism doing something, along with the details and all the accessories that also had to be included," Jomard wrote.

Conté planned far more ambitious projects that were never implemented. Before the Abukir disaster, for example, Conté had proposed that the French establish a telegraphic system that could warn the French admiral at the enemy's first appearance. The army ignored his advice. After the disaster, Conté established furnaces in the Alexandria lighthouse to heat and smelt cannon balls to repel an English attack on the city from the sea. The English did not even attempt to land, though, hewing to a strategy of letting the Turks and disease do proxy battle for them.

At an especially dark moment in the occupation of Egypt, when the last of the original French-made tools had been destroyed in a battle, Napoleon indulged in an unusual moment of despair. "What are we going to do now? We do not even have tools!" the general cried, according to Jomard, who recounted the incident: "Conté, who was present, replied, '*Well, we will make the tools.*' And he did what he had said."

CHAPTER 4

THE INSTITUTE

It was once a country to be admired; now it is one to be studied.
— From the entry on Egypt, *Encyclopédie*
(Diderot and d'Alembert)

*O Egyptians! You have been told that I have come to this land
only with the intention of eradicating your religion. But that
is a clear lie. I, more than any Mameluke, worship God, glory
be to Him, and respect his Prophet and the great Koran.*
— Napoleon Bonaparte, in an announcement
read in Arabic to the people of Egypt

Cairo, Late Summer and Fall, 1798

Shafts of early morning sunlight angle in through the *mashrabiyah*, spraying tiny geometric shadows across the floor and walls. The intricately tooled mahogany window screens are designed to let rare breezes in, while allowing cloistered women a surreptitious view out onto the street. On this morning, however, at seven A.M., the twenty-second of August, 1798, no women are present, nor any Arabs at all, only fifteen French scientists, already sweating inside their frock coats, and General Napoleon Bonaparte and his guards.

Gaspard Monge strides to the front of the room, silently noting the tiny squares and circles of dark and light around his feet. The shapes are pleasing to the geometer's eye; he is, after all, the inventor of a mathematical technique to calculate and represent the throwing of shadows. The perfect angles of these shadows are perhaps a good omen for the first meeting of the Institute of Egypt (or, Institut d'Égypte). Down one floor, fountains burble in the interior courtyard, long parlors link a labyrinth of cool rooms, and orange blossoms scent the walled gardens. Everything within this compound is in peaceful symmetry, antithesis to the clamor on Cairo's streets.

The scientists' meeting room is a high-ceilinged interior parlor, lapis-tiled and furnished with silk velvet and damask-covered daybeds and chairs, and rare potted plants. A faint odor of gardenia and musk permeates the air. The room was, until a few weeks ago, when the Mamelukes fled the approaching French, a harem parlor, province of indolent, perfumed women and their eunuchs.

Reaching the improvised podium, the geometer greets his colleagues. Only fifteen are present because most of the Institute's members have still not arrived in Cairo. Monge hails this memorable day, the inauguration of a body whose goal is to "foster the spread of enlightenment and knowledge in Egypt." Then he introduces the man who has made it all possible, General Bonaparte.

The first order of business is electing leaders and there is only one obvious leader in the room. Bonaparte, however, declines the presidency, and the honor goes to Monge. The scientists make the general their vice president. Fourier, still in Rosetta, is elected in absentia to be the Institute's permanent secretary.

"Citizen Bonaparte" now steps forward and addresses the scientists, humbly thanking them for the honor of membership, and predicting great victories for knowledge ahead. He then poses the first six questions for study in Egypt. One: Could the ovens used for baking army bread be improved? Two: Could they make beer without hops in Egypt? Three: How might Nile water be purified? Four: Which would be more practical in Cairo—windmills or watermills? Five: Could gunpowder be made in Egypt? Six: What is the state of civil Egypt, in terms of law and education? What do the citizens want?

The scientists in the room—mathematicians, naturalists, astronomers—nod and agree they'll get right to those useful tasks. They are each thinking, though, of many other matters that they'd like to study: mirages, for one, about which Monge is prepared to read a paper. Or: How to move a monumental stone bust of Isis? How to preserve a crocodile skeleton? How best to dissect bird mummies? Why does sel ammonium form naturally in the desert? Can ostriches fly, and if not, why do they have wings? What causes the endemic eye disease temporarily blinding soldier and scientist alike? And what is the meaning behind the colossal ruins?

The first meeting is adjourned long before noon. The members of the newly minted Institute make their way back out into the streets, where braying donkeys, water-laden camels, French soldiers, and, mostly, Cairenes of all colors and sorts—Africans, Albanians, Arabs, Greeks, Muslim holy men (some clothed, some naked), the blind, the feeble, the strong, the beautiful, and the grotesque—push, jostle, and struggle in a synthesis of sound, smells, sights, and dry heat the likes of which the French have never experienced.

A l-Qahira. The Arabic name for Cairo means "the city victorious." Cairo was indeed victorious in its resilience, a lasting human hive on the edge of vast desert, but when the French arrived, it was the city mysterious: a labyrinthine metropolis that frustrated and confused the intruders. They found a city of doors, mostly closed. Massive gates opened into the city, and the winding streets themselves often ended abruptly at smaller doors that defined neighborhood and community boundaries. In Muslim urban life, the private and public realms were rigorously divided, and the French crossed these lines at their peril.

Women lived mainly indoors, appearing on the streets enveloped in veils, as they do today. The definition of the private domain was larger then, however. Whole neighborhoods might be walled off, accessible only by a single door in a narrow street. These doors inconvenienced the French, and eventually Napoleon committed one of his most offensive acts—in the eyes of the Arabs—when he ordered them removed.

When the French arrived, Cairo (called Fustat by the first Arabic invaders in the seventh century) had existed as a human settlement for more than a thousand years, and the history of human habitation in the area, attested to by the Great Pyramid at Giza, stretched back far beyond what Napoleon's scientists reckoned of the ancient world. It was one of the world's largest and most cosmopolitan cities, a warren of twisting, dusty, narrow streets teeming with animals and people from Asia, Africa, and Europe. In the Middle Ages, half a million inhabitants made Cairo the biggest city the Western world had seen since ancient Rome. In 1798, a quarter of a million people lived in Cairo, and the city was still an independent, thriving center of commerce. For centuries, Cairo had served as a trade crossroads between East and West. Enormous caravans of slave traders, pilgrims en route to Mecca, as well as silk, spice, and gold merchants passed through the city year-round. Scores of Arabic words for the goods that passed through Cairo have enriched the English language. In the realm of cloth alone, our word for cotton comes from the Arabic word qutn, gauze comes from Gaza, "dimity" cotton from the name of the Egyptian port city Damietta.

Travelers approaching Cairo from the desert on camel or foot first glimpsed the hazy spires of hundreds of minarets, rising from a flat yellow plain, and in the foreground, bluish hints of triangles, as the pyramids appear first on the horizon. They entered Cairo by one of seventy-one towering gates, and made their way inward on lanes dominated on one side by the towering pale-gold cliffs of the Mokattam Hills, ancient source of rock for the pyramids. Massive cemeteries in the city center—the Cities of the Dead—covered hundreds of acres, with narrow lanes winding between domed tombs and minarets holding the remains of a thousand years of Muslim faithful, from average families to ninth-century mystics and great sheiks. Living people inhabited this vast quarter, too, the caretakers of the tombs, whose jobs were passed down through the generations. The massive, white, crenellated walls of the Citadel, a military fortress built by the Mamelukes in the Middle Ages, towered over the eastern edge of the city. Between these landmarks was a web of footpaths and alleys, barely wide enough for two donkeys to pass, lined with houses and shops packed so tightly together the blinding noon sun rarely penetrated to street level.

Cairenes initially panicked and feared the French invaders, and surrendered almost immediately. They seemed docile enough, but their submission was temporary. They were certainly not fooled by Napoleon's insistence on his adherence to the Muslim faith. Muezzins called prayers five times daily from the minarets of hundreds of mosques, in a language the French, with the exception of a few translators in the civilian corps, did not understand. Soon enough, these mournful-sounding prayer-callers would turn against the occupiers.

The confounding layout, the narrowness of the streets, the doors, the unfamiliar language, all combined to make the city very difficult for the French to truly conquer. The city frustrated Europeans. To their eyes, there was no logic to its street plan, and less order. Claustrophobic alleys ended at walls, or dwindled into walkways and disappeared. Camel, donkey, and human feet vied for space on these dusty ways. Donkeys knew the city better than the French, and soon the conquerors had requisitioned the humble little beasts for their own uses, straddling them so their boots dragged on the ground and treating them like horses, beating the sleepy creatures with horsewhips to speed them up.

Some French, like Geoffroy Saint-Hilaire and Denon, fell in love with Cairo at first sight, while others—many in the military—never saw its charms. Most were simply amazed at the juxtaposition of filth and beauty, ruination and construction, richness and poverty. Geoffroy Saint-Hilaire thought the soldiers hated Cairo because they were mostly idle there, having nothing but water to drink and only veiled women to ogle. The garbage and dirt also had something to do with their unhappiness. Hills of refuse, hundreds of feet high and centuries old, rimmed the city. Ordure lay in heaps along the streets, too. The French—soldier and scholar alike—never failed to mention these stinking piles in their letters home.

To the Arabs, though, it was the French who were rank. The contemporary Egyptian historian Jabarti noted that the French soon ordered the people "to sweep, splash water, and clean the streets of rubbish, filth and dead cats." Jabarti found this ironic, considering the French were so filthy themselves. "This [order] was in spite of the fact that the streets and houses where the French lived were full of filth, infected earth mixed with bird feathers, the entrails of animals, garbage, the stench of their drinks, the sourness of their alcoholic beverages, their urine and excrement, such that a passer-by was obliged to hold his nose."

Beyond the garbage and disorientation, the newly arrived French found Cairo painfully loud. The young director of the printing press, the Orientalist Jean-Jacques Marcel, was housed near a troop of dervishes, and he wrote that between their shouts and the barking of the city's wild dogs, he lived in Milton's Pandemonium. The barking dogs so annoyed Napoleon that he soon ordered his soldiers to kill them all in one night. The stray dogs were so numerous that the canine massacre, carried out with sabers, took two nights instead of one.

The French, coming from a country that had just banished religion and proclaimed secular humanism the state faith, arrived in a city animated by fervent religiosity. Cairo regularly erupted in saints' festivals, or *moulads*, and the French could barely begin to fathom their meaning. August 1798 coincided with two of the country's holiest days, the

Festival of the Nile, which had pagan roots in ancient Egypt, and the birthday of the Prophet Muhammad. When the Institute held its inaugural meeting, the Festival of the Nile was in full swing.

The festival began on the day every summer when the river rose to its annual flood stage, which was always heralded by the return of the sacred ibis. At that point, religious leaders ritualistically ruptured the dike and flooded the city's canals, turning the city into an African Venice for a month. People then picnicked by the waters and celebrated all night in the illuminated streets. The elegant Ezbekiyah Square (where Napoleon had requisitioned himself a Mameluke palace), near the city center, was turned into a lake during the flood. Illuminated boats appeared, and celebrants threw candy and coins to children and the poor on shore.

The Nile festival was followed a few days later by the Prophet's birthday, occasion for an even wilder public party, with gladiators, performing monkeys, dancing bears, street poets, and dervishes and naked holy men doing ecstatic dances. Napoleon seized the opportunity of the festival to win Egyptian hearts and minds by pouring money into the celebration and thereby portraying the French as true Muslims. "Inhabitants of Egypt!" his first proclamation began. "The French adore the Supreme Being, and honor the Prophet and his holy Koran! The French are true Muslims. Not long since, they marched to Rome and overthrew the Pope, who excited the Christians against the professors of Islamism." Napoleon signed the document as "Bonaparte, member of the National Institute, Commander in Chief."

The remaining *savants* reached Cairo as the celebration reached the height of frenzy. Many of them, like the engineer and Institute member Étienne Malus, were utterly baffled by what they saw. "I arrived at 11 o'clock in the evening. All the streets were illuminated. One sees in the public places people ranged in a circle, shouting repeatedly, falling, spitting, reaching extremities of fatigue and rage. These are the saints of the country. Their life is a continual ecstasy; all is permitted them. Many run in the streets naked as monkeys. They live off charity."

The evening after their Institute's inaugural meeting, some of the scholars joined the crowds celebrating the birth of Muhammad in Ezbekiyah Square. The French military participated by firing cannons, in order, Jomard wrote, "to prove to the inhabitants that, far from wanting to annihilate their culture, we want to take part, as it were, in it." Whether the Egyptians interpreted the cannon fire as celebratory or threatening, Jomard did not think to speculate. Some Egyptians did acknowledge the French presence by serving tricolored rice at feasts.

Monge and Berthollet had gone to Cairo early in search of the perfect space in which to house the Scientific Commission's members and their laboratories, tools, and collections. They had their pick of any number of the city's fantastic mansions, left undefended by the Mameluke warriors who, if they had survived the Battle of the Pyramids, had retreated to the desert. Though abandoned, many of these mansions still housed a harem of women and eunuchs. What the scientists wanted was a compound big enough to house a library and museum, chemistry and physics labs, an observatory, a museum, a meeting room, workshops, a printing press, and a zoo. In other words, nothing short of a palace would do.

Nestled behind high walls away from stench and din, Cairo contained pockets of luxury—Mameluke splendor—which the French sought out and claimed for themselves. Napoleon took a palace on the elegant Ezbekiyah Square, and his top lieutenants were equally well housed in other recently abandoned Mameluke mansions. The rank and file merely seized the homes of the better class of Egyptians.

After a few days surveying Cairo on donkey-back—the preferred mode of transport in Cairo, the wheel not much used there before 1800—the two scientists located the real estate of their dreams. A group of empty, contiguous palaces built by wealthy Mamelukes offered luxurious sleeping quarters, vast meeting halls, and a walled garden that covered thirty French *arpents*, the *arpent* being roughly equivalent to an acre. The palaces were more than a mile from the general's headquarters, a problem only when the scientists needed military protection—which, as it turned out, would happen sooner than expected.

Geoffroy Saint-Hilaire, arriving early from Rosetta, also participated in these initial arrangements. The young zoologist was ecstatic at the choice of housing, and, for that matter, thrilled by Cairo itself. "Things are even better in Cairo than in Rosetta," he mused to his friend, the natural-history painter Henri-Joseph Redouté, who was still in Alexandria. "Our lot has already improved. Magnificent lodgings; immense and beautifully drawn gardens; abundant and flowing water softly murmuring from every side; a multitude of different species of trees offering voluptuous shade; the society of all the generals and particularly that of the General-in-Chief." He compared the commandeered palace and its furnishings to the Louvre in terms of magnificence. All he lacked, Geoffroy Saint-Hilaire wrote, was a competent manservant who could groom a horse and cook his meals. He urged Redouté to find him one, and to get to Cairo as soon as possible, since the rooms were being parceled out and already the first arrivals had snagged the best of them.

To his father Geoffroy Saint-Hilaire wrote that even the sight of French soldiers riding Cairo's donkeys, their long legs grazing the dusty alleys, charmed him. "It is very funny to see all kinds of Frenchmen meeting one another while riding donkeys," he wrote, adding that while it was easy enough to control the donkeys, the reaction of the human Cairenes to the French invaders was a different story. "The men seem indifferent, but the women are scared and always cry and scream that the French want to force them to change religion. However, the women are beginning to calm down."

Within days, Geoffroy Saint-Hilaire was collecting curious animals from all over Cairo—ostriches, birds, snakes—left in the empty Mameluke houses, and bringing them together in his garden zoo. He had already erected an aviary, and protested vociferously when the Egyptian house servants ate one of his rare birds before realizing its significance. He boasted in letters home that the Egyptian Institute's growing animal collection would very soon be better established than the one at the Jardin des Plantes.

Beyond the Institute's walls, the seemingly chaotic Al-Qahira was organized in ways that would take the scholars and French military months to understand. Water, for example, was daily brought into the city from the Nile on the backs of men, donkeys, and camels, who together were supervised by a guild. Another organization oversaw the donkey taxis. Professional street cleaners watered and swept the streets, although many of them had fled the city when the French invaded, leaving no one to provide these and other essential services. A guild of lamplighters lit the streets every night. As the French invaded, these basic services were disrupted, and the French administration had to struggle to understand how they worked and how to get them running again.

Much of the city's civic services were owned and operated by religious foundations called *aqwaf*, and the final arbiters of law and order were the *ulema*, or sheikhs, whose authority was based on Islamic tradition. Religious foundations operated the *sabils*, or public water fountains—elegant, rounded structures carved with calligraphy and adorned with marble columns. The *sabils* housed cisterns at street level, but also religious schools for orphans on the second floor. In a city without running water, these public fountains allowed human life to proceed, and their dual religious and civil use made sense.

To the Arabs, who knew its culture and history intimately, Cairo was beautiful, "the mother of cities," in the words of one fourteenth-century Arab author. The scholar Ibn Khaldun called Cairo "the metropolis of the universe, garden of the world, swarming core of the human species, a city embellished with castles and palaces, bedecked with convents and colleges, illuminated by the moons and stars of knowledge." The French scholars soon learned to seek out the city's serene surprises, the hidden gardens scented with orange and lemon trees, shaded by hanging vines, acacia and myrtle, giant-leaved banana, pomegranate and mulberry trees.

Even the confusing streets were part of a grand plan that reflected a type of thinking completely different from the European notion of urban design. The modern Egyptian novelist Gamal al-Ghitani has written that the endless turns in the streets had a psychological purpose. The city's architects broke them up purposely, he believes, to create *wa'ad al bisul*—"promises of arrival." A street that veers sideways, in this view, momentarily provides the weary traveler with a feeling of relief, of being somewhere, that he would not experience walking on a long, straight road.

Like Paris, Cairo in 1798 had a lively café culture. The French scholars counted 1,350 coffeehouses in Cairo. Outside these cafés, professional poets, who wore rushes on their heads as identification of their profession, recited extemporaneous verses on the streets. Unlike Parisian cafés, these little coffeehouses, to the chagrin of the French, didn't serve wine. Cairenes had no need for alcohol or tobacco, having refined their own preferred intoxicants long before the Europeans arrived. A special guild sold honeyed hashish, and the coffeehouses sold opium. The scholars, in their eventual report on the mores of modern Egyptians, estimated that the poorer people used drugs all the time, in their homes and at work.

When the French arrived, various European-style vendors suddenly appeared. Cairo had a large population of ethnic Albanians, Greeks, and Muslim Slavs from the European edges of the Ottoman Empire. These entrepreneurs knew just what the homesick French missed most. They set up shops catering to all the French vices, purveying tobacco and wine and women. The most extravagant pleasure palace opened in November 1798, a Cairo version of Tivoli Garden, a park with music and refreshments, for the exclusive amusement of the French. Here, Napoleon first laid eyes on the fetching French lieutenant's wife whom he would make his Cairo mistress. Napoleon could dispatch his new lover's lieutenant husband to France because, although the French had no fleet left, merchant ships still sailed between Egypt and Europe, and the French were able to board them for a price. They did not always get home in a safe or timely manner, as Dolomieu and Dumas, among others, discovered.

Napoleon was besotted with his own wife Joséphine, but a few months into the Egyptian campaign, he learned via letters from his family that she was being unfaithful to him back in Paris. After a period of dramatic bereavement, in which he tore at his hair, languished in his tent, and told alarmed battle companions he would never recover from the heartbreak, he rallied and took up with a lieutenant's wife named Pauline Fourès. Madame Fourès was an intrepid charmer better known as "Bellilotte," a nickname derived from her maiden name, Bellisle. Napoleon conveniently dispatched her husband to France on some pretext and set her up in a mansion next door to his on Ezbekiyah Square. Thus favored, the pale-skinned Bellilotte earned the nickname Cleopatra for the rest of the campaign. She returned to France in 1800, remarried (having divorced the lieutenant while still in Egypt), and went on to live a long and colorful life, even writing a novel and becoming a prosperous merchant in Brazil, from where she returned to Paris an older woman with a collection of parrots and monkeys, who shared her apartment. She died in 1869, with the Suez Canal under construction, having survived her Egyptian lover by more than four decades.

By September, most of the scientists and artists had arrived in Cairo, and the Institute's business was well under way. The Institute's founding document named three purposes, in an order that pleased the scientists, enlightenment and study preceding the government's more pedestrian needs. The three official goals were:

"The advancement and propagation of enlightenment in Egypt.

"The research, study and publication of industrial, historical and natural phenomena in Egypt.

"To offer its opinion on different questions on which it will be consulted by the government."

The scholars had taken over three contiguous Mameluke mansions, and were now housed in the utmost luxury. Savigny and Geoffroy Saint-Hilaire, along with a few other scholars, shared a sumptuous tiled palace, where they went to sleep every night to the sound of a fountain in the courtyard. In the garden of the compound's main building, the scholars erected a giant sundial, bearing the inscription "L'AN VII RF" (i.e., year seven of the revolutionary calendar of the French Republic).

Here, the scientists began organizing their Institute and its offices. Although much of their equipment now lay at the bottom of the Mediterranean Sea, they had been able—with Conté's help—to replace some of what had been lost. Besides a large library culled from their own private collections, they installed chemistry and physics laboratories, eventually erected an observatory, a natural history exhibition room, and a small cluster of mechanical workshops, where Conté and his staff worked. Among the outbuildings and gardens, Geoffroy Saint-Hilaire's aviary and zoo (with squawking monkeys, plus an Egyptian mongoose, gazelles, and snakes) were joined by a small archaeological museum that eventually housed various French finds, including the Rosetta Stone.

The Institute was modeled after the prestigious French Institute (Institut de France) in Paris, and it was selective, with only fifty-one elected members, although all civilians and military men were invited to attend meetings and contribute reports. Members were divided into four areas, two of them scientific ("Mathematics" and "Physics") and two in the humanities ("Arts and Letters" and "Political Economy").

Nonmembers could, and did, submit reports on everything from the flora and fauna to Egyptian religious practices and superstitions, myths, and curiosities they encountered in Egypt. None of the students were elected Institute members, but they were integral to the scientific community. One of them, the young botanist Ernest Coquebert de Montbret, eventually became the Institute's librarian, a post he held until his death, at the end of the expedition.

The *savants* tried to fulfill Napoleon's more practical needs even as they worked in their own fields. At the fifth session they reported that certain Egyptian materials, including reeds, were better than wood for heating ovens. They looked into the possibility of growing grapes for wine, and compared the properties of Egyptian wheat with French wheat. Desgenettes conducted a study of hospitals. Monge and Berthollet were put in charge of the Cairo mint, and served on a commission that extracted money from wealthy Cairenes. Fourier was put in charge of Egypt's first newspaper, *Le Courrier d 'Égypte*. At the Institute's tenth meeting, Napoleon submitted a new list of queries to his human encyclopedia. His requests included studying the Nilometer (the ancient Egyptian device for measuring the rise and fall of the Nile), how to carry water to the Citadel, where the military was headquartered, the feasibility of an astronomical observatory (they erected one), and a study of aqueducts.

The scholars also tried to accomplish the Institute's first purpose, "bringing enlightenment to Egypt"—with mixed results. Only a few Egyptians, of the higher classes, spoke French. The entire community was welcome at their public lectures about chemistry, electricity, botany, morphology, and other subjects from the scholars' individual fields. Typical of the response was one theologian, who after listening to Geoffroy Saint-Hilaire describe the fish of the Nile, rose and announced that such research was pure vanity since the Prophet had declared that God had created 30,000 species, 10,000 on land and 20,000 on water.

Despite being literally stranded in Egypt, most members of the Commission were optimistic during their first weeks at Cairo. They felt they had created a "scientific city." Cairo was distant from Paris, it was chaotic, and many men suffered from dysentery and fevers and homesickness, but the climate was strangely conducive to intellectual work. Geoffroy Saint-Hilaire was in transports, comparing the Institute's fragrant gardens to those of the Jardin des Plantes, while noting that he lived in a *"foyer des lumières"* (home of enlightenment).

"Here I once again find men who think of nothing but science; I live at the center of a flaming core of reason," Geoffroy Saint-Hilaire wrote to his father in October. "We busy ourselves enthusiastically with all the questions that are of concern to the government and with the sciences to which we have devoted ourselves freely."

Geoffroy Saint-Hilaire was not alone in that heady sensation. "We have on the other side of the palace, the vast garden of Kassim Bey [a prominent Mameluke in hiding] to walk in during the evening, and the charming conversation of Fourier," wrote Jomard. "The beauty of the sky, the perfume of oranges, the softness of the temperature, all add more agreeableness to these gatherings, sometimes prolonged into the middle of the night. It is our garden of Academus. More great thoughts, more true philosophy, more scientific discoveries are being born, and we flatter ourselves that we are laying the foundations of a new school of Alexandria, which will one day efface the old."

Jomard also praised the after-dinner salons, which he found to be especially stimulating. "Besides the regular sessions of the Institute, informal gatherings of forty or fifty people took place every evening in the garden of the Institute," he wrote. "They would talk about their travel projects, about discoveries they had made, and about a variety of fascinating questions concerning the physical geography of Egypt, ancient Egypt, the government of the country, and the mores of its people."

One reason for the sense of intellectual excitement was the mixed nature of the scientific community. Science then was far less specialized than it is today, and highly compartmentalized fields of study did not yet exist. Back in Paris, though, there were boundaries between general fields; chemists were not expected to theorize about zoology, botanists were not conversant in mathematical theory, for example. Parisian science also adhered to rigorous standards of proof and theory. In Cairo, the boundaries between fields blurred, and Parisian rigor diminished. In the Institute gardens, architects debated with naturalists about animals and ancient structures, physicians and astronomers debated with the geographers about the meaning of the hieroglyphic script, the age of the ancient culture. These conversations among learned men manifested the highest ideals of the Enlightenment. Especially for the younger participants, the effect was of embarking on a thrilling intellectual exercise without borders.

The frequent presence of Napoleon himself heightened the intoxication. "I owe Providence great actions of grace for all the blessings with which I am showered," Geoffroy Saint-Hilaire enthused in a letter home. "I enjoy more affluence than I had in Paris; I have the happiness to approach our illustrious leader and to eat with him rather frequently. His actions have caused him to be proclaimed, deservedly, the man of the century; his intimate conversation taught me that he was the best of men."

In these early months in Cairo, an eclectic spirit of inquiry animated the Institute's regular meetings. Scientists read reports on lunar cycles, Egyptian music, and the physical properties of recently excavated and dissected mummies. Scientists at the September 12th meeting, for example, discussed the baking of bread in Egypt, a series of algebraic equations, the establishment of a journal, a translation of an Arabic book, and chief medical officer René Desgenettes's report on the endemic maladies of Egypt, including dysentery and ophthalmia.

There was, of course, the not insignificant problem that this "flaming core of reason" was well out of range of Western civilization. As conceived back in Paris, the Egyptian Institute was supposed to liaise with the French Institute at home. The *savants* addressed their reports to the mother academy, but quite soon that gesture became only symbolic. The scientists did manage to send occasional texts back to

France, but after the destruction of the fleet, and with the English in control of the seas, Paris and Cairo rarely communicated. The English began intercepting most French correspondence, which they then published, with sarcastic annotations. The lack of word from home severely diminished French morale over time. The English soon had an effective propaganda campaign disseminating throughout Europe letters written by homesick and lonely Frenchmen in Egypt.

A few letters of instruction arrived from the Institute in Paris before the English cut off all correspondence between France and Egypt. In one, the astronomer Pierre-Simon de Laplace asked his colleagues in Egypt to perform certain astronomical observations. Mostly, though, the letters from Paris just never arrived, and the Institute was left to its own devices.

The *savants'* own correspondence that was seized by the English was a great source of amusement to the enemy. The satirist James Gillray drew cartoons of hapless French scholars being chased by crocodiles, or pinned by furious Arabs and Turks at the top of Pompey's Pillar, instruments and notebooks flying through the air. English antiquaries like William Hamilton, however, were keenly interested in what the scholars were up to, as the French discovered, rather unhappily, a few years later.

The Institute proceeded with its work, blissfully unaware of its laughingstock image in Europe. With its own printing press it was soon putting out two publications that kept scholars and soldiers alike abreast of doings within the Institute and throughout greater Egypt. Every ten days, the journal *La Décade Égyptienne* was published with selections from the reports read at the Institute's meetings. A newspaper, *Le Courrier*, published more frequently, shared news and gossip from within Egypt (a typical issue reported on a caravan of one thousand Nubians arriving in Cairo and a recent battle against the Mamelukes in Upper Egypt) and, whenever available, news from Europe.

These two publications are historic because they were the first documents, besides Napoleon's initial decrees, ever printed in Egypt.

The Mamelukes' religious leaders had tightly controlled information and forbidden the use of printing presses, believing that manuscripts needed to be transmitted by hand and analyzed individually to preserve their power and integrity.

The Institute kept copious notes of its meetings, but its records disappeared in France sometime in the early nineteenth century. *La Décade*, though, did survive the years. These reports show the Institute often hewed to scientific standards of proof, but sometimes strayed into what we might call popular science. Members and nonmembers read reports about their travels beyond Cairo, with casual observations on ostriches, Arabian horses and the Egyptian fly, the Nile lemon, topographical studies, translations of Arabic poetry, the curative powers of olive oil for plague victims, and, later, Geoffroy Saint-Hilaire's musings on "anatomical philosophy."

The meetings could be lively, if esoteric. In fall 1798, the scientists engaged in a three-hour debate—sparked by a report from Geoffroy Saint-Hilaire—over whether the ostrich could actually fly. Ostriches were ubiquitous in Egypt, and so many soldiers had taken up ostrich hunting that within months of the invasion, it was rare to see a soldier without a few ostrich feathers jauntily decorating his hat. A member of the French military happened to be present at this extended argument at the Institute. His outraged report to fellow soldiers on the triviality of the scholars' interests—debating the purpose of the giant bird's wings!—swelled the resentment soldiers felt for the scientists.

In the heat and dust of Cairo, relations between soldiers and scholars never improved. The soldiers nicknamed the city's ubiquitous donkeys *"demi-savants,"* a reference to a story that, during an early battle, Napoleon had ordered his soldiers to form squares with "donkeys and *savants* in the center" for protection. "To calm the jealousy of the soldiers, Bonaparte openly joked about the *savants*," the engineering student Jollois noted in his journal, adding this anecdote: " 'They closely resemble women, no?' Napoleon said one day to [chief medical officer] Desgenettes. 'General, one has a bit more fun with women!' said Desgenettes. 'Oh! But I was talking about their whisperings and their rivalries and their pretension!' "

The scientists were mostly impervious to the soldiers' gibes. They truly were the general's favorites, protected, coddled, and consumed with a sense of discovery, confident of the importance of their work. In one letter to Cuvier about his plan to send regular Institute reports back to Paris, Geoffroy Saint-Hilaire wrote, "I assure you that these meetings are at least as interesting as those at the Institut de France."

Beyond the Institute walls, Napoleon was having little success winning over Egyptians, though not for want of effort. He was portraying himself and the French soldiers as liberators; unfortunately, however, the cultural and religious divide was proving too wide for his rhetoric to bridge. The Arabs preferred their Mameluke overlords to European invaders. Napoleon's defeat of the Pope was no selling point, either. In the Muslim world, infidel Christians were a notch above godless atheists.

Napoleon made some inroads with those Egyptian leaders who had acquiesced early (motivated, no doubt, by one proclamation's threat to "burn down the villages" that did not raise the French tricolor). The ultimate authorities over civil life in Mameluke Cairo were religious teachers and jurists, called the *ulema*. These law sheiks were in charge of "ordering good and denying evil," and Napoleon organized them into the first representative assembly in the Middle East.

In early September, Napoleon created a General Divan consisting of deputations from each region, including law sheiks, merchants, Bedouin, and peasant. Monge and Berthollet attended as its French commissioners, and the body met just once a year for two weeks in early October and regularly throughout the occupation. Napoleon asked the Divan to discuss how best to organize Egypt, including issues regarding provincial assemblies, a criminal justice system, and what improvements could be made in the existing structure of taxation and property ownership.

The timing was right for reforms. Egypt was feudal, but the Mameluke landowners (for whom ninety percent of Egyptians toiled) were on the run, and socialists among Napoleon's advisors were recommending a wholesale redistribution of property, with every peasant given some land. Other advisors, more conservative, advised keeping the old feudal system intact and extracting acreage for deserving Frenchmen instead. In the end, neither reform was ever officially brought before the Divan.

The Divan experiment did not augur well for the goal of bringing democracy to the Egyptians in 1798. The assemblage, which included the scholar and chronicler Jabarti, had to be taught how to elect a chairman because none of them had ever voted before. After two weeks of deliberation, the group recommended that nothing at all be changed in the existing way of life. The Divan did continue to meet throughout the occupation, though, and served as a means of communication between French leaders and the Egyptian people—mainly relaying news of increased taxes.

Napoleon also tried his luck with Egypt's religious leaders, hoping, as he put it, to "lull fanaticism to sleep before we can uproot it." To do this, he told the theologians that he and his army wished to become Muslims. The plan fell apart when the religious leaders demanded circumcision and abstention from alcohol.

Napoleon's conversion was more likely to amuse or offend the Egyptians than win them over. He dressed in "Turkish" costume—pantaloons and turban—on at least one occasion, but his advisers thought he looked so ridiculous they begged him to change back into his own clothes before stepping outside. Napoleon later denied donning the outfit, although he did insist he would have happily professed faith in the Prophet and worn the turban—and ordered his troops to do likewise—if it had seemed politically expedient. "Will it be said that the subjugation of all Asia were not worth a turban and a pair of trousers?" he asked his comrades years later in exile. In the end, Egypt didn't require mass circumcision, or the change of clothes.

Napoleon was always proud of his fraudulent religious gambit, even though at best it merely baffled the Arabs. He never regretted it. "This was quackery, but it was quackery of the highest order," he boasted late in life. "Change of religion for private interest is inexcusable. But it may be pardoned in consideration of immense political results."

Napoleon won some admirers, especially among the poorest Egyptians, who called him *Sultan Kebir*, Great Leader. In Egypt, Napoleon felt closer to what he thought was his true destiny as a military and spiritual leader than he did at any other time in his career. He was especially pleased when he could take the opportunity to posture Solomonically. Years later, he shared with his courtiers a favorite memory of one such incident. While visiting a small village he learned that Bedouin Arabs had killed an innocent peasant. Napoleon publicly vowed to chase the perpetrator's tribe into extinction in the desert. The local sheiks laughed and asked, "Is the dead man your cousin?" Napoleon replied, "He was more than my cousin; all those I govern are my children. Power is given to me only that I may ensure their safety." On hearing these words, Napoleon recalled, the sheiks all bowed their heads and said, "Oh, that is very fine. You have spoken like the Prophet."

The scholars fared better with the Arabs than did the general and the French army because they tended to associate with the Arab civilians more as equals, or at least as consumers, rather than invaders. They hired *fellahin* (peasants) to perform household work, bought food and horses and equipment from Arab merchants, and interacted respectfully with the religious scholars.

The *savants* often invited Cairo's religious leaders to the Institute, where Berthollet and Geoffroy Saint-Hilaire entertained them with experiments in electricity, chemistry, and dissection. The Egyptians often sat impassively before the scientific magic, though they reacted with horror when they were administered small electric shocks. The scholars were disappointed by the Arab disinterest, but they didn't give up. If the Arabs were bored by French mastery of the basic forces, surely they would be agog at human flight. In his first months in Cairo, Conté made two hot-air balloons and sent them aloft with much fanfare, hoping to impress the population. The first balloon was launched on the occasion of the anniversary of the French Revolution, a month after the French arrived at Cairo. The balloon was large—twelve meters in diameter—but it was made of paper, which soon ripped and caught fire. When the Egyptians saw the whole contraption coming down in flames, they ran for their lives, convinced that the balloon was a war machine.

Conté was undeterred. He launched a second hot-air balloon in the center of Cairo in December 1798 to mark the anniversary of the Battle of Rivoli, one of Napoleon's successes in Italy. This balloon was made of canvas and even larger than the first. It rose, floated, and landed without incident, but this time the inhabitants, from the most educated religious men to the poorest *fellah*, showed no surprise. Some people even crossed the square without looking up at the balloon once.

Jabarti recorded the Arab opinion on this particular display of "Frankish" science: "What they promised did not come true. They claimed a sort of ship would travel through the air thanks to the marvels of technology. In fact it was no more than a kite such as menservants make on holidays and at public festivities."

The Arab theologians didn't show it, but Jabarti conceded that they did admire the scientists' methods, if not their materialism and extravagant claims. "We were.shown other experiments as well," he wrote, describing displays of electricity and chemical processes, "all as extraordinary as the first one, such as intelligences like ours can neither conceive of nor explain."

There was a strong, albeit long-lost, intellectual tradition in Islam familiar to scholars from both cultures, but European science had long since eclipsed Arab learning. Arab science had protected classical knowledge through the European Dark Ages, between the seventh and twelfth centuries, when plague and war reduced Europeans to subsistence and brutishness. During that period, the Arabs translated and improved upon Greek theories, and produced many great thinkers, including the medical philosopher Avicenna (or, Ibn Sina) and the scientist Al-Razi, who laid the foundation for modern chemistry in his laboratory. They were accomplished mathematicians; it is not for nothing that the numerals we use today are, in fact, Arabic. For a time, astronomers were essential members of the courts in Baghdad and Cairo. These men were actually all-purpose intellectuals, practicing astrology and astronomy as a combined science, and sometimes writing poetry as well. In the West, the poet Omar Khayyám is the best-known of these astronomer-poets.

By 1798, however, Arab science had long since fallen into decline, reaching a low point in which only religious texts were deemed worthy of serious study. The cadre of European scientists alarmed the religious learned men of Cairo but also piqued their curiosity. Jabarti had a low opinion of the French in general, but he was impressed by the scientists' library and their generous, open attitude with regards to knowledge. "The French installed a great library, with several librarians who kept guard over the books and handed them to those readers who needed them," Jabarti wrote. "When a Muslim wished to visit the establishment, he was not prevented from doing so, but on the contrary, he was made very welcome. The French were particularly pleased when a Muslim visitor showed an interest in the sciences. I myself repeatedly had occasion to visit the library."

Napoleon, twenty years later, wrote that the Egyptians came to appreciate the *savants*. "The native population [at first] thought that they were making gold." Gradually, though, as the scientists worked with both the notables and the working people, teaching them elements of mechanics and chemistry in the process of helping build roads and make other civic improvements, the Egyptians came to hold the *savants* "in high esteem."

Cairenes could appear blasé about European science and technology, but they were openly shocked and offended by some of the French customs, and none more so than the Frenchmen's attitude toward women. This particular aspect of the culture clash was unprecedented and especially disturbing to both sides. Through the centuries, Islam had developed a very restrictive attitude toward women, among whom modesty was prized above all else. Even the most influential Muslim women were always veiled and secluded. When the French arrived in Egypt, among their numbers were several hundred officers' wives and female camp followers, women who went about freely in public with men as they wished, and who never covered their heads. French women rode around Cairo on donkeys, dressed in revealing garments, laughing and talking with whomever they happened to meet—Arab or French, male or female. Some Arab women began to imitate the European manner of dress and behavior, but their show of independence lasted only as long as the French controlled Egypt. As soon as the French left, their religious leaders publicly beheaded some of these rebellious souls as examples to their wayward sisters.

The French, for their part, were astounded to find that despite all their modesty, Cairenes, especially during festivals, could be positively lewd in public. The French military chronicler Bernoyer wrote of watching a woman riding through the city totally naked on a magnificent mare. Even though she appeared to be around sixty years old, and her breasts hung to the saddle, Bernoyer found her rather attractive. "Her figure was distinguished, and young, she must have been very beautiful," he wrote. "Her gray hair matched the gray hair of the horse, that was slowly led by a slave holding the bridle." Bernoyer watched in befuddlement and amazement as people approached the naked woman one by one, and touched her buttocks with their fingertips, in attitudes of profound reverence. "Without my interpreter, I could not learn the virtue of this woman's buttocks!"

In their final book, the scholars reported on what they termed the Egyptians' *"bizarre jouissances"*—strange sensual pleasures. The young engineer Jomard witnessed an even more prurient scene involving one of the so-called *santons*—Muslim mystics. "Everything is permitted to them," he wrote in *The Description of Egypt*: "They can do whatever they want. One *santon* who was supposedly inspired by Mohammed, had the habit of walking around in the city entirely naked. Women who saw him did not turn away, but instead stopped and kissed his hand." Jomard heard tell of one *santon* who took a woman and "overturned" her (*renverser*—possibly Jomard's euphemism for having sex) in the middle of a populated street. Another woman passed, took off her veil and covered the couple. The first woman then harangued the people herself and said that the Prophet's inspiration itself had brought the saintly man there, and that there would be born from their union a faithful believer.

Toward the end of October, the illusion of peaceful coexistence between French and Egyptians dissolved on a single morning. Enraged at various French demands, in particular an onerous new tax levied to make up for the fortune lost at Abukir Bay, Cairo revolted, in a citywide uprising instigated by the city's religious leaders. The epicenter of the insurgency was the massive Al Azhar mosque and theological center, the black and white marble bastion built in 968. With its sprawling area and three minarets—one for Shi'a, one for Sunni, and one for peace between the two Muslim sects—the compound dominates central Cairo to this day. Three thousand students and sheiks studied in this sanctuary with its elegant rows of columns and blue-tiled prayer nooks, whose intricate geometric patterns uncannily suggest an a priori knowledge of particle physics.

On the morning of the twenty-first of October, acting on word from the theologians, the muezzins began calling from a hundred minarets across the city, urging the faithful to take up arms and repel the French. As one observer, a writer known as Nicholas the Turk, recounted, the French were oblivious to the population's submerged rage, but the theologians were not. "One fine day, some sheik or other of Al Azhar started to run through the streets shouting, 'Let all those who believe there is but one God take themselves to the Mosque Al Azhar! For today is the day to fight the Infidel.'" Within minutes, crowds were running through the streets, brandishing sticks and swords, in scenes that surely reminded some French of Paris in 1789. Even so, Napoleon reacted with initial nonchalance. After putting his troops on alert, he left the city to inspect some fortifications. Two hours later, he returned to find one of his generals shot and corpses piling up in the streets.

The scholars' Institute was a mile from Napoleon's palatial headquarters, and the scientists were left with no choice but to mount their own defense. As a mob gathered outside the compound's walls, chanting and throwing rocks, the scholars inside bickered over the best course to follow. Some saw an opportunity to escape and wanted to run, but Monge took charge and shamed them. "Would you dare to abandon the instruments of science entrusted to our care?" he bellowed. No one dared.

Inside, more than a hundred panicking civilians, utterly ignorant of military strategy, prepared to defend themselves and their scientific instruments. They took up knives and sticks, and tore up the terrace, so as to hurl tiles at their attackers. Pandemonium passed for defensive strategy. "Each had his own plan, but no one would obey anyone else," Denon later wrote. Some pleaded with Dolomieu, as a former Knight of Malta, to take command, but he refused the role.

A small group of engineers lodged at a contiguous palace were killed within hours. The mob broke a wall and massacred the inhabitants, including the chief cartographer in charge of mapping Egypt, D. Testevuide (none of the primary documents seems to give his full first name), who had rushed into the building to protect astronomical and mathematical instruments, which the looters smashed. They then fed the engineers' corpses to the (apparently surviving) street dogs, after demolishing a huge cache of scientific equipment stored inside.

The insurrection lasted two days and nights, but the scientists managed to protect themselves and their compound. As soon as he was able, Napoleon sent an aide-de-camp to the scholars to make sure they were surviving, and armed them with forty rifles and 1,200 cartridges.

Two hundred French soldiers died before the army crushed the rebellion with a massive bombardment of Cairo and Al Azhar, killing 2,000 Egyptians in the process. "This bombardment was so terrible that the inhabitants of the city, who had never seen such a thing, began to cry and pray Heaven to preserve them," wrote Jabarti. "The rebels ceased fire, but the French continued the shelling. Houses, shops, palaces, inns—everything crumbled. One's ears were deafened by the sound of the guns. People left their houses and the streets to hide in holes."

The first French soldier to break into the Al Azhar compound and crush the core of the insurgency was General Thomas-Alexandre Dumas, who charged into the sanctuary astride a horse whose nostrils spurted blood, swinging a saber above his head, looking like a biblical apparition. "The angel! The angel!" shouted the fleeing Arabs, according to Dumas's novelist son, Alexandre.

General Dumas, the mixed-race bastard son of a French marquis and his Caribbean mistress, was born in Santo Domingo and became one of Napoleon's greatest warriors. Black-haired and with a sculpted, Herculean physique, he inspired legends with his feats. He sometimes amused his friends by setting four infantry muskets on the floor, inserting a finger into the barrel of each, and raising them simultaneously to shoulder height. As a young man in riding school, he supposedly could stand up in the stirrups, take hold of an overhead beam, and lift himself *and* his horse bodily off the ground.

Dumas was a passionate fighter, one of the first men off the ships at the stormy Alexandria landing. Arriving on terra firma, he immediately borrowed a musket and charged into the Egyptian interior with a small band of intrepid followers. He later grew disgusted with the failures of Napoleon's Egyptian war planning and was sent back to France early, on a ship that wrecked near Taranto, Italy, where he was captured and imprisoned for months. When Dumas finally returned to France, his military commission was never renewed.

Once his soldiers restored order in Cairo, Napoleon punished a few dozen surviving insurgents with public beheadings. Many French, even the humanist Denon, complained that the general was too lenient in his assignment of guilt and that more Egyptians should have been executed, as a lesson to future rebellion-minded mullahs. In fact, the insurgency was never really extinguished, and a larger, longer, and more damaging insurrection lay ahead.

After the insurrection, French soldiers pillaged and desecrated Al Azhar, a profound offense to the Egyptians and to Muslims in general. This vast black and white striped marble structure, whose compound enclosed acres within the city, was not only a great mosque, it was the oldest and most venerated religious educational institution in all of Islam. As part of this revenge, Napoleon even allowed his horses to be stabled inside the mosque courtyard, a profound and unforgivable offense. Jabarti described the final outrage: "They treated the books and Koranic volumes as trash, throwing them on the ground, stamping on them with their feet and shoes. They soiled the mosque, blowing their spit in it, pissing and defecating in it. They guzzled wine and smashed their bottles in the central court." Jabarti called them the "host of Satan."

In the aftermath, the scholars from the *foyer des lumières* could do little but form a clean-up crew. A group of scientists went to the mosque, rescuing some manuscripts, including an ancient Koran inscribed entirely on camel-skin, from the wreckage. It is unclear whether they returned them to the rightful owners or increased their own collection of curiosities. Geoffroy Saint-Hilaire admitted that he acquired for himself, in this foray, an antique, annotated Koran.

CHAPTER 5

THE ENGINEERS

> We were many times obliged to replace our weapons with geo-
> metrical instruments, and in a sense, to fight over or to conquer
> the terrain that we were to measure.
> —Joseph Fourier, *The Description of Egypt*

Egypt, Fall and Winter 1798–1799

Berthollet recruited the bulk of his younger scholars—thirty-six men—from the École Polytechnique. When word
of the mystery expedition leaked through the elite school late in the
winter of 1798, students fought to be picked. It was, wrote one of
the young engineers who signed on, "an epidemic madness," fueled
by hero worship of Napoleon Bonaparte. The students who were
eventually selected were the cream of the crop, hand-picked by their
professors, some of whom were also on the expedition.

The École Polytechnique contributed a full third of the Commission members, including faculty. Although it taught hard science,
the school encouraged independence of thought and was considered
experimental for its time. Students chose the classes that they wanted
to take, and had the choice of doing or not doing the mathematical
problems and chemistry exercises proposed to them every month.
They were rigorously trained in drafting, and especially in descriptive

89

geometry, the new way of applying numbers to the world discovered by Monge. The professors, chosen (like Monge) from the ranks of the nation's top mathematicians and chemists, were under orders to teach by Socratic method, rather than droning at a room full of dozing boys.

Students came from all over France, and boarded with families in Paris, who were instructed to keep them away from the city's iniquitous corners. Paris life was stinted, food was rationed, and bakeries were often empty. The students developed an *esprit de corps* and a sense of healthy competition. They came mostly from well-to-do families, but the streets of postrevolutionary Paris were rough, and the boys learned to use their fists as well as their drafting pencils. Intramural brigades regularly joined in combat on the school grounds, and students brawled outside the school with soldiers and street gangs.

These young men, ranging in age from sixteen to twenty when they sailed, were billeted with the lowest-ranked soldiers, in the most crowded holds of the fleet's least luxurious ships. For thirty days, belowdecks, the young students slept on layered hammocks that reeked of tar and leather, with soldiers above, below, and beside them. They were ill fed, subject to soldiers' taunts, unprotected, and unsupervised.

The student-scientists were nominally under the command of the army's commander of engineers, General Maximilien de Caffarelli du Falga. Caffarelli, as he was known, was a war hero with a wooden leg and an intellectual cast of mind. He had fought with Napoleon in Italy and was one of the young general's most trusted military companions. During one of Napoleon's shipboard salons, the talk turned to a discussion about ideal forms of government, and in an extemporaneous speech, Caffarelli offered a detailed proposal to abolish private property. Historians of the journey later compared it to a fully imagined pre-Marxian socialist system that was never put to pen. Like some of Napoleon's other officers, Caffarelli genuinely respected the *savants*.

He also befriended and protected the younger men. Once the students landed in Egypt, they actually found the battle-scarred general a more attentive mentor and guide than their somewhat chilly young professor, Joseph Fourier.

Most of the students already knew each other from Paris, but they formed tighter bonds on the ships. Some were bookish and obedient, priggish and copious note-takers; others were independent, even unruly, and sowing wild oats. Like their elders, all had been molded by the Revolution. The studious Édouard Devilliers du Terrage (Devilliers to his friends), son of aristocrats, had spent his teenage years, the bloodiest years of the Terror, hiding in a Paris attic with his younger sister, selling off the family library and silver, piece by piece, as his father was imprisoned and his mother died. After the Revolution, he managed to enter the École Polytechnique, where he heard about the mysterious expedition, signed up, and was accepted.

Devilliers's friend and fellow student, Jean-Marie-Joseph-Aymé Dubois (he called himself Dubois-Aymé on the expedition to distinguish himself from the five or six other Duboises in the civilian corps), was a muscular, passionate, athletic young man familiar with hazards and thrills long before he joined the mystery expedition. His family had noble ancestors and was linked to the royal family when the Revolution came. Dubois-Aymé's father administered the king's farms until shortly after the king was executed, when he fled the country with his family. Dubois-Aymé crept back into France as a teenager. After a period in hiding, he reemerged in Paris and registered at the École Polytechnique, where he became known for his willingness to engage in fisticuffs, and his winning way with women.

The two eighteen-year-olds shared a coach from Paris with their professors. True to character, Dubois-Aymé was diverted by a little romance before he boarded his ship. At Marseille, he seduced an Italian beauty. This woman, whose name has been lost to history, was already mistress of the aforementioned General Dumas, reputed to be "the most good-looking man in the army" and "the strongest man in the army." Dubois-Aymé dallied with Dumas's fetching girlfriend for a few days, arriving at Toulon later than his student comrades and missing his assigned boat.

Dubois-Aymé was blasé about the beginning of the great adventure. He kept an intermittent journal, noting that he found shipboard life "rather boring," and the accommodations cramped and miserable. "We fight, we grumble, we feel rather sick, we drink bad water, and we do not have a place to run, or even a tiny room in which to be alone for a single instant," he complained. To escape the stinking, cramped quarters, Dubois-Aymé spent most of his time strapped high in the rigging, reading and gnawing on biscuit and cheese.

In a closet-sized berth he shared with twenty-four soldiers, his friend Devilliers was not nearly as footloose. Profoundly seasick, Devilliers dutifully made daily notes in his journal between retching episodes, chronicling the cruelty and discomfort of military life. As the ships set sail, he witnessed two sailors summarily shot for trying to escape by jumping overboard and swimming to shore. Reeling with the horror of that scene, he crept back to his hammock, only to have it break under his weight, dropping five feet to the floor. Within days a soldier stole the rest of his bedding, and Devilliers was obliged to find a new place to sleep. He chose a heap of coiled rope on deck.

Devilliers's best friend, Jean-Baptiste Prosper Jollois, didn't even bother telling his family he was leaving until he arrived at Toulon. Just before sailing, he sent a letter informing his father of the alarming fact that he was embarking on an expedition to an unknown destination. Jollois, twenty, had a remarkably philosophical attitude toward life and fate. He told his father that even as the ships lifted anchor, he remained ignorant of the destination, but that he was trusting his leader and his luck. "One must hope that the Government will not abuse the blind confidence that a large number of people have in it," he wrote, adding that he was prepared for adversity. "Whatever may arrive, I am awaiting the worst, as it is the means of always being agreeably surprised."

Some students were homesick before the boats even lifted anchor. Seventeen-year-old botany student Ernest Coquebert de Montbret, a boy who hadn't begun to shave, wrote to his parents from Marseille that he had cried on reading their last letter to him. "Your sensitive and tender feelings were painted in such a vivid manner that I was unable to restrain my tears."

The young man idled away the hours, trying to be inconspicuous in his assigned hammock, sandwiched between, above, and below hundreds of sweating, seasick soldiers who, when they were not vomiting, amused themselves teasing the smooth-cheeked boy. Coquebert bore it stoically, complaining only of the monotony. "All my companions suffer cruelly from boredom and idleness. We don't lack books, but we can't read all day." The students tried chess and lotto, but the games soon became "insipid and long." At least, he wrote, the food on his boat was good.

That last item was a short-lived luxury. Shipboard conditions deteriorated rapidly for the low-ranking soldiers—an ominous early sign of the poor planning behind the general's mystery expedition. Some boats faced food shortages within weeks. Soldiers soon began trading their clothes for rations. Devilliers wrote that by the second or third week, the lamb and other fresh meat on the boats was rotting and, almost worse, being served up raw. Fresh water was scarce and reserved for drinking. Within days of embarkation the odor of unwashed flesh was inescapable.

"We are 110 in a room that is 30 square feet, and with what society! What terrible racket!" complained Devilliers in his journal. "The food is mediocre, but what is disagreeable to the highest degree is the water that one must drink, black and smelly enough to make you step back."

The students cringed at the soldiers' shipboard amusements belowdecks. Soldiers played cards, sang obscene songs, and invented prurient skits that, as the secret destination became known, invariably included a beautiful slave girl trapped in a harem by an old Turk, guarded by eunuchs, needing to be liberated, then married, by a French soldier.

The soldiers also alleviated their own shipboard ennui by teasing the young scientists. To the military rank and file, the scientists were less useful than women. Clumsy and seasick, the civilians were hopeless as deckhands. The soldiers resented the pudgy, soft men taking up valuable hammock space. Early in the sea voyage, one general publicly suggested that he wanted to throw the useless scholars overboard. Whenever they could, soldiers took the scientists' bedding and rooms by force. They were especially enraged that the younger professors—twenty-one-year-old naturalist Savigny, for example—had high rank, shared tables with officers, and spent evenings with the general in chief. Napoleon ignored the complaints. He gave his *savants* military rank and pay (they declined uniforms).

Soldiers took to calling the scientific corps *"la maîtresse favorite du général"*—the general's favorite mistress. That was among the kinder epithets the fighting men coined for the scientists in the months and years to come.

By the time the ships reached Malta, the younger scholars were sick of the mystery, the abuse, the deprivations, and the mention of the trip itself. Some contemplated going AWOL. For the randy Dubois-Aymé, at least, the notoriously beautiful, dark-eyed Maltese women, not to mention the platoons of streetwalkers in its city, were a powerful inducement to remain on the island as long as possible. After three weeks on the water, the young men were quite happy to linger in their luxurious Maltese quarters, a huge hall, fifty feet long, decorated with silver-framed portraits of the Knights, and stocked with fresh oranges and good wine. Dubois-Aymé and the boy botanist Coquebert actually hatched a plot to stay in Malta and ultimately make their way back to France. Devilliers, however, would not provide aid and encouragement.

"I will try hard to change Dubois's mind," Devilliers wrote in his journal from Malta. "If he insists, I will be quite angry, but I will tell him *adieu* and feel sorry for him. In effect, he will have done a great chore to come to Malta and then return to Paris, with the shame of not wanting to continue."

Devilliers eventually persuaded his young friends to get back on the boats, partly by painting a fantastic picture of Egypt. "We suspect the goal of the voyage, and if it is truly what we believe, it could only be very advantageous," he wrote in late June. "As for me, even if I were to be ten times more badly off, I would not quit. I will not stop even if I should meet a ditch in the middle of my race."

The young engineering students arrived at Cairo a day before the first meeting of the Institute of Egypt. Spared the desert march, they traveled via the Nile on one of the small sailing ships that had plied the waters from the delta to the cataracts for millennia. Mosquito-ravaged, sometimes feverish or gut-cramped, the students veered between exhilaration, boredom, and despair. Some days they were thrilled to be in Egypt, on other days, homesickness seized them, or they were overwhelmed with insecurity and confusion about what they were supposed to be doing in this furnace of a country.

When the young men entered Cairo, they had already been in Egypt for six weeks and were accustomed to life on the Nile. They had observed the river's human, insect, and animal inhabitants and bathed in its waters for relief from the scorching daytime heat. One member of their group had already drowned in a river whirlpool at Rosetta.

"I swam across the Nile, it is about 600 meters wide," Devilliers wrote in his journal at Rosetta. "The water is muddy, and one is stung by little fish armed with fins that end in very sharp points." The river's banks, he noted, swarmed with scorpions, and the Egyptians had taught the French to remedy the painful sting with the burning application of "a red iron," the student wrote. The occasional scorpion, though, was nothing compared to the ubiquitous clouds of mosquitoes and flies. Devilliers wrote that so many mosquitoes had attacked him and his colleagues that they all looked like they had the pox. "There are twelve flies on my hand as I write this," he finished.

Having witnessed the thousands of French bodies washed up on the sand at Abukir, the young men had learned some things about death in Egypt. They had also tested the limits of their own endurance. Some had been very sick themselves. They'd slept outdoors, gone without food and water, been attacked by insects, dogs, and men. They were still nominally students, but they had learned one lesson all too well: none of their Paris training in measuring elevations, drafting architectural plans, and surveying topography had prepared them for life in Egypt.

Devilliers first caught sight of Cairo while he was still suffering from a lingering fever—which he had initially feared was plague—contracted in Rosetta. He was simultaneously revolted and intrigued by the city, as this entry in his journal the day after arriving in Cairo shows: "The façades of the houses are original. There are balconies with pretty wooden grills that enclose them. Most have baths. Rugs or mats cover the floors. Most of the houses are ruined and most of the streets are disgusting, especially where they dry manure that the Arabs use as a combustible. The strangest streets are those where the stores are. The storekeepers sit outside their stores, drinking coffee and smoking pipes all day long."

Prosper Jollois was pleasantly surprised by his first sight of Cairo. His rooms on the Institute's palatial grounds cheered him up even more. "I admit the repugnance I had in coming to Cairo, and the fear that I would be totally abandoned in a city not vaunted for its beauty dissipated little by little at the sight, of which the aspect appeared very agreeable," he recorded primly in mid-September.

The young men entered Cairo at the height of the festivities celebrating the Nile flood and the birth of the Prophet. On their first night in the illuminated city they were met by a cacophony of snake charmers, performing monkeys, braying donkeys, jugglers, belly dancers, dervishes, and street poets. "Everyone is out at nine at night, singing and doing bizarre dances," Devilliers wrote. "The songs are supposed to imitate the cries of a mother and child. A dozen dancers stay in a line by holding hands and are preceded by a *santon*. They hold each other around the chest tight enough to suffocate one another, they sway, screaming, kissing the spectators' hands, giving their hands to be kissed, and look as if they are saying their last goodbyes. They continue like that until one of them drops, exhausted, and often dead." Devilliers was mesmerized by the *santons*, whom he described as "sort of crazy men who are very venerated and who are allowed to do whatever they want. Their insults are considered honors, and they even insult women."

At first, the young men didn't have much to do in Cairo but soak up the street scene. Devilliers's journal is filled with entries that read only, *"rien de nouveau"*—nothing new. To earn spending money, he took a job. Geoffroy Saint-Hilaire assigned him to collect ostriches from the Mameluke mansions commandeered by the various French officials. He captured one for the Institute's zoo, but his second foray came to naught. "I was supposed to get another ostrich from the Pasha's house but the people of the house had eaten it in their master's absence," he noted on August 26.

The Cairo streets fascinated the students, but as budding engineers who had spent several years studying the design and construction of buildings and bridges, what they really wanted was to inspect the oldest man-made structures in the world, shimmering in the haze on the horizon at Giza. From a distance, the pyramids are translucent blue triangles against the yellowish sky, ephemeral as dreams. The dusty air and their own elegant angles make them seem to hover above the ground. Only on closer inspection do they become what they really are, massive, man-made mountains of rough-hewn, square rocks.

The engineering students could not wait to get close to these monuments to ancient engineering. The only way from Cairo to Giza was by boat, and special arrangements had to be made with the French military for any such visits. Prosper Jollois was one of the first students to go. Naturally, his instinct was to ascend, but climbing was not as easy as it looked. "I arrived halfway up the pyramid with incredible difficulty. The tiers, almost worn by the time and encumbered with the detritus of the upper stones, did not at all allow one to place one's foot in security, and more than once, I walked on an inclined plane that collapsed under my steps. That, added to stones that those who walked above me made fall, and that were flung with such force they could have knocked over the man they would have hit, convinced me to turn back."

Eventually Jollois clambered to the top via a more stable route. He also inspected the face of the Sphinx nearby, noting that an entire body was "said to be under the sand."

Devilliers and Dubois-Aymé also visited the pyramids, crashing the party of a more senior group of scholars. Monge and Berthollet arranged an outing for themselves and other luminaries of the scholarly Commission. Students weren't invited, but the two young men, hearing of it, crossed the Nile the night before and hid, fully dressed, in the nearby Giza palace. In the morning they were first in line for the boat to the pyramids. They had already settled in when the older scholars boarded. The scientific leaders were surprised to see the students, but no one reproached them.

On this trip, Napoleon dared Monge to race to the top of the Great Pyramid in the burning September sun. The ever-game geometer, fortified with gulps from a bottle of brandy he always carried, beat the younger men to the peak. After surveying the Nile valley from this emotional and physical height, Monge wrote his wife to predict that Europeans would someday winter in Egypt, and that the French would have themselves an earthly paradise there in fifty years.

A few days later, Devilliers had another opportunity to admire the uncanny imperturbability of his elder colleagues, when he picked up a gun in an Institute room with Berthollet and Geoffroy Saint-Hilaire. Completely unaware the weapon was charged and ready to fire, he accidentally shot a ball between the heads of the other two men, breaking a mirror behind them. "Neither one nor the other made any response!" wrote the astonished student.

Joseph Fourier was nominally the students' leader, but the thirty-year-old professor had a distant, selfish personality and had already left his young scholars in dire straits on several occasions in Rosetta and Alexandria. The students had learned not to count on his help when they were in need, but to seek out other men. The younger men took solace from Berthollet's example of quiet solidity. "During his stay in Egypt, Berthollet was insensible to privations and fatigue," Jomard wrote in a biography of the chemist. "Happy with everything, complaining about nothing, one would have thought that he was enjoying all of the comforts of life in Europe. His simplicity was extreme. He was affable and obliging to the young chemists on the expedition."

Caffarelli, socialist, engineer, survivor of numerous battles—and with a wooden leg to show for his service—was another man to whom the students turned when their spirits flagged or they wanted supplies. "When I find myself sad, beaten, a half-hour conversation with him suffices to put me right," the young botanist Coquebert de Montbret wrote to his parents.

The soldiers, though, resented the chief engineer. They believed Caffarelli had been one of the original promoters of what they increasingly viewed as the doomed Egyptian adventure. Whenever he limped past, someone would mutter, loud enough for him to hear, "He doesn't care what happens; after all, he has one leg in France."

In early September, Caffarelli ordered the youngest engineering students, those who had not yet graduated from the École Polytechnique, to prepare to take their final exams in a few weeks. The students took out their books and pencils and went to work. The Institute's grounds—with Geoffroy Saint-Hilaire's diverting menagerie of ostriches and mongooses—were not ideal for studious concentration, but for serenity, the compound was unmatched in Cairo. They lined up their chairs, faced the wall of books in the library, and bent over their calculus, trigonometry, and drafting.

On October 6—almost the same date on which their classmates back in Paris were taking their final exams—the students stood one by one before a committee comprised of Monge, Fourier, and Berthollet. They all passed. Graduation did not confer membership in the Institute, but it did raise their status and put them on a more equal footing with the soldiers and the younger professors. From then on, the students participated in every aspect of the Institute, attended meetings, dined and socialized with the scholars and high-level military officers on the Institute grounds.

As soon as they passed their test, the graduates were asked to decide which branch of the Commission they wanted to join, the military engineers or the corps of bridges and roads. Napoleon needed these adolescents and young men for their vigor, spirit, and rigorous training. With a little direction from their elders, they could already conduct mapping surveys and build roads, design bridges, locate and route fresh water into canals. If the young men eventually found other aspects of Egypt more interesting—and most of them would—they would never be allowed to forget that they had jobs to do for the French military first.

The newly minted engineers were put to work in two general areas: mapping and hydrography. The only available map of Egypt in 1798 was a seriously flawed effort drawn in 1765 by the chevalier d'Anville. Described as an "armchair cartographer," d'Anville never set foot in

Egypt; he worked entirely from books and older maps. During the first year of the expedition, the engineers raced to produce more accurate maps for the military. Lacking the latest equipment (lost at sea), they relied on older, more fallible surveying methods, including compasses and measuring chains, which they confirmed, when possible, by more sophisticated astronomical calculations.

The mappers owed what accuracy they achieved to the astronomer-priest and Commission member Father Nouet, who had kept his own instruments, including marine chronometer, repeating circle, and telescope, in his personal baggage. Using his gear, Nouet was able to assist the mapmakers in ascertaining astronomical points corresponding to geographical locations.

Given the civilians' limitations, in terms of gear and mobility, the maps they eventually produced were, not surprisingly, substandard. In the final book containing the maps, Colonel Pierre Jacotin, who directed the mapping work in Egypt, apologized for their crudeness, blaming the lack of precision instruments, the limited number of men to conduct the work, and the shortage of military escorts to guard them.

The mapping process went like this: an engineer worked on a portion of terrain corresponding to a series of astronomical points graphed out by astronomers. Walking from one astronomical point to the next, he measured the distance between the points with a chain. Every one or two thousand meters, he marked his progress on a piece of paper with a small measured stick called a *graphomètre*. Once he arrived at a new point, he repeated the process, proceeding to the next astronomical point.

What the mapping project lacked in precision engineering, it made up for in descriptive science. Along with their measuring equipment, the surveyors carried forms on which they were instructed to record detailed information about the people, animals, and plants they encountered while working. Each form contained ten columns, with slots for place names in French and phonetic Arabic, numbers of family members, their work status, the type of agriculture in the region, local industry, commerce, even names of the trees. The notebook was for recording more general observations about water and air quality, indigenous animals, and local character, tribal movements, customs, and anything else the surveyor might consider pertinent.

This dual demographic/land survey process was almost as dangerous as it was tedious. It took three months to map Alexandria alone, mainly because of the extreme summer heat. Mappers ran from wild dogs and armed insurgents, brushed away scorpions and flies, and suffered thirst, sand in their instruments, and blinding eye infections.

The first engineer in charge of mapping Egypt, Testevuide, had been murdered during the Cairo insurrection. His replacement, Pierre Jacotin, was ordered to conduct a detailed survey of the teeming metropolis, on a scale of 1 to 2,000. The Cairo mapping job was deemed so daunting that at first the engineers hoped the order would be rescinded, but it was not. Finally, Jomard, accompanied by a few Egyptian interpreters who could speak Arabic and read Coptic, ventured into Cairo's winding lanes. For weeks he traversed the city by donkey with his guides. He started by assembling—in Arabic and French—all the names of the streets, places, monuments, markets, and public establishments in the city, while also collecting information on industry and commerce. Working in withering heat from sunup to sundown, he rapidly became more familiar with the doors, alleys, canal beds, and arches of Cairo than most of the French civilians would become in three years.

Jomard produced a full report in less than three months, noting in his preface that he had been especially challenged by the doors. "The city is almost entirely composed of very short streets and twisting alleys, with innumerable dead-ends," he reported. "Each of these sections is closed by a gate, which the inhabitants open when they wish; as a result the interior of Cairo is very difficult to know."

Jomard did his best to know it, though. On his mapping forays around the city, he noted aspects of Cairene culture far beyond
place names and topography. Wandering the narrow alleys with his
donkey and interpreters, knocking on gates that hid whole neighborhoods, he scribbled reminders in his notebook about encounters with
snake charmers, belly dancers, and religious mystics, and of children
glimpsed through doorways of the madrassas, busy memorizing the
Koran, a method of teaching that Jomard suggested in his reports
might be superior to the French way.

"In one public place, singers gather and play wind and string instruments," he reported. "There are also illusionists who perform for the
public, who do tricks similar to those that illusionists in France do, but
they also perform other ones that French illusionists would not do. For
example, they cut off the nose of a child and produce an illusion that
is so cruel that one involuntarily steps back when the mutilated child,
its face bloody, comes to ask for some money. These same illusionists do tricks with trained monkeys, and they play with scorpions and
snakes."

Jomard had not led an especially sheltered life back in Paris, but
the extreme poverty and licentious adult entertainment he observed
in Cairo still shocked him. He found the provocative *a'lmeh* practicing
their art on many streets. These belly dancers usually entertained
inside the harems and the houses of the rich, but they also worked
the streets—at least, he surmised, "the most vulgar ones." The *a'lmeh*
danced in public every day except during Ramadan, and Jomard was
astonished by the openly sexual nature of the show. "If two *a'lmeh*
dance together, one plays the lover and they enact scenes that are not
at all subtle and that are full of licentious gestures." He recorded and
translated some of the salacious lyrics, including, "Come, my friend,
and undo the cords of your belt and approach me."

Jomard was the first scholar to systematically inspect the homes
of Cairo's poorest people, the *fellahin*. He initially thought the dugout-

style dwellings he passed were doghouses, since they were only four feet high, constructed of clay mixed with stones and open on top. He soon learned otherwise. "It is hard to believe that humans can actually live there, since these places are so low and small," he reported to his colleagues later. "An entire family can live in holes that are six feet in diameter. The misery and dirtiness of these people make one step back with disgust." The blackened interior walls were proof that the inhabitants lit fires inside. After seeing dogs, goats, and sheep poking their heads out of third-story windows, he realized that domestic animals also shared quarters with the better classes of human families as well.

In the end, Jomard and the corps of engineers produced many sheets of topographical maps of Cairo and the rest of Egypt. Flawed though they were, the French government deemed them so valuable they withheld their release to the public for nearly thirty years. The caution and pride attached to these documents probably reflects a lingering French delusion about victory in Egypt. The published forty-seven-sheet topographical map of Egypt was substandard for the day, with inaccurate measurements and erroneous assumptions about property lines and cultivation. The *savants'* maps were nevertheless the best available for Egypt in the first half of the nineteenth century. They gave future travelers and mappers at least a reasonable outline of the country's geography, without letting them forget who did the hard work. When the mapmakers published their maps of Egypt, each plate included, in the upper-left-hand corner, the distance of the represented place from the meridian of Paris.

Besides mapping, the young engineers were assigned another, equally important, job: the detection, use, and management of water. When the French arrived, the now-dammed Nile River was still its own master—annually flooding to levels dictated by nature, not man. Fed by monsoon rains in Africa, the river rose and crested in late summer, overflowing its banks, then receded, leaving fertile muck for the short growing season. Long months of dust followed, as people, plants, and animals lived off the harvest and waited for the next inundation.

In the thousands of years of recorded history before the first Aswan Dam was built in 1902, the Nile flood regulated Egyptian life. People marked time by it, named gods to rule the river, celebrated festivals around its cycles. Even the device used to measure the water level was sacramental, and its care was entrusted to priests. When the French arrived, the Nilometer in Cairo, on the Nile island of Roda, was located at the end of a long avenue of sycamores, surrounded by a grove of lemon and orange trees.

The river's spiritual significance even migrated into ancient Greek and Roman mythology, and touched European Christianity as well. Aristotle reported on the Nile's many supposed magical properties. He wrote that Nile water boiled in half the time it took other water. He also credited the Nile for the fecundity of Egyptian women, who supposedly often gave birth to twins, and after just eight months. Nile water was even exported to ancient Rome for religious purposes. The Vatican itself contains a symbolic marble sculptural group depicting sixteen children, one for each ell of the Nile flood, playing at the feet of the river god. "The Nile surpasses all rivers of the inhabited world in its benefactions to humanity," wrote the first-century-B.C. historian Diodorus Siculus. Seneca wrote that compared to it, all other waters were *vulgares aquae.* Egyptians were believed to dislike wine because the water of the Nile was so sweet.

After enduring the forced march through the burning desert to Cairo, the French easily understood why the Egyptians had worshipped their main source of fresh water. No French soldier could argue with the river's life-giving capacity.

Besides figuring out how to irrigate the desert, the French badly wanted to answer another crucial question about Egyptian waterways. Two great seas lapped Egypt's shores, tantalizingly close to one another, separated only by about one hundred miles of desert. If a canal could be built linking these seas across the top of the African continent, boats would no longer have to sail around Africa to get to Asia, and merchants trading with India wouldn't have to hazard the perilous land trek. The implications of a canal for colonial control of the East were enormous.

A canal was not a new idea. Geologists believe the isthmus between the Red Sea and the Mediterranean was still a strait fairly recently in the planet's history and that alkaline lakes in the desert between the two seas are remnants of a prehistoric waterway. Over the years, the two seas receded and river channels diminished as the area silted up. The process of silting up was probably well under way during the pharaohs' time. The ancient Egyptians dug many canals to keep the Nile delta open for boat passage, but they never actually connected the two seas via a hundred-mile-long ditch. At best, they had linked the waters of the Nile and the Red Sea, via trenches through the desert to the port of Suez.

Egyptians did build one great canal in the seventh century B.C. According to Herodotus, 120,000 men died digging a ditch said to be wide enough for two armies abreast. Invading Persians added to it centuries later, but left it unfinished because they believed the Red Sea was higher than the Mediterranean and feared that a canal linking the seas would inundate the freshwater Nile with salt water.

Europeans had been interested in an Egyptian canal for centuries, as an easier East-West commercial route than rounding the Cape of Good Hope. In the seventeenth century, a French merchant, Jacques Savary (not to be confused with the travel writer Savary), mapped out details of possible Egyptian canals and calculated how much they would cost. He even drew up precise plans for a canal between the Red Sea port of Suez and the Mediterranean, and during the reign of Louis XIV opening the Egyptian isthmus became a matter of French national policy. In 1665, the French invited the Turks to help build a canal, hoping that the promise of increased revenue from customs duties would persuade the Sublime Porte. The Turks were too antagonistic toward the Christian West to consider any kind of joint venture. What's more, they restricted parts of the Red Sea to Muslim use, owing to their proximity to holy sites. For the next century they continued to use the same reasoning to veto every new French argument in favor of building a canal.

In the late eighteenth century, the French desire for a canal was sharpened by the quest for a larger role in international commerce. Pilgrims to Mecca had to cross the Gulf of Suez, and the Turks had by now completely closed the Red Sea to the French above Jedda, explaining that infidel sailors might desecrate the Prophet's tomb. In 1798, French merchants operated in the lower Red Sea within these religious restrictions. They hauled coffee from Arabia and cloth from India across the water to southern desert ports, then packed their wares across the desert along old caravan routes, and finally floated down the Nile to the Mediterranean. The British had forged a special agreement with the Mameluke beys to allow their goods from India to travel all the way up the Red Sea to Suez. By the time Napoleon was dispatched to Egypt, the French badly needed an Egyptian canal to aid their own colonial ambitions—and to humble England's.

Napoleon had not forgotten the French government's last directive: "He will then cut the Isthmus of Suez and take all necessary measures in order to assure the free and exclusive possession of the Red Sea for the French Republic."

On a chilly Christmas Eve, 1798, he set off on a personal scouting mission with his favorite scientists, some engineers, and a contingent of guards and headed east into the desert from Cairo. Monge, Caffarelli, and Jacques-Marie LePère, the engineer who would eventually be charged with surveying a possible canal, rode up front with the general. Various Cairo merchants affixed themselves to the rear of the armed French caravan, seeing it as an unusually secure method of reaching the Red Sea. With this entourage of nearly three hundred men, the general set off.

A few miles east of Cairo, the terrain turned into a bone-dry moonscape of gravel and sand, devoid of life except for the occasional ostrich, gazelle, and bird of prey. No speck of vegetation offered food or fuel. A single sacred acacia tree marked the pilgrim station of Haura in the desert. Its branches festooned with bits of clothing left by pilgrims en route to Mecca, Napoleon took the precaution of sleeping under the tree himself to prevent his soldiers from cutting it up for fuel. To warm themselves that Christmas Eve, soldiers coaxed fires from animal bones scavenged from the surrounding wastes, and their Yule feast of dry bread and biscuit and water sucked from goatskin canteens was better suited for ascetic monks than the avowed atheists they were. Napoleon and Monge, however, somehow contrived to dine on roasted chicken.

After a few days' march, the caravan reached the derelict port of Suez. Only a few Arab dhows were docked in the silted harbor, but Napoleon saw the potential in what he described as "a squalid and filthy place." He ordered the harbor dredged and boats sent from Cairo. Late in the afternoon, at low tide, he led a group across the peninsula to drink from the biblical Wells of Moses, with Monge providing a running discourse, complete with quotes from Exodus, on the flight of the Jews across the Red Sea and their wanderings in the Sinai.

The gang nearly perished later that night. The Red Sea rose faster than expected and the Arab guides, warmed with French eau-de-vie, got lost in the gathering dusk. Men and beasts began sinking in the watery sand, and the sea soon engulfed the horses' bellies. Caffarelli, mentor of the student engineers, lost his wooden leg in the mire and nearly drowned before a soldier rescued him. Guards watching the expedition via spyglass from land recognized the trouble their general was in and set a house on fire to guide the entourage back to shore as night fell. All were saved (except, presumably, for the Egyptian's house). Napoleon later honored Caffarelli's rescuer with a sword inscribed with the legend "The Crossing of the Red Sea."

On the return to Cairo, Napoleon was the first to sight the vestiges of the thousand-year-old canal banks in the shifting sand of a dry wadi. The trip was deemed a great success, though on January 7, the student Jollois—who was not in the group—noted caustically in his journal: "Everyone who had been to Suez returned. The General in Chief made those who accompanied him suffer a lot by excursions and races. He killed two or three horses, lost a guide, and nothing slowed his ardor."

When the Institute met in January, an engineer read a very tedious report on Egyptian agriculture, an Orientalist provided a French translation of an Arabic poem. When the discussion turned to work assignments, many of the young engineers understood for the first time that they were in Egypt to do dangerous fieldwork of special importance to the military. A number of them were reluctant to cooperate, but they were given no choice in the matter.

In a particularly bitter journal entry, Jollois denounced the officer in whom he and the other young scholars had placed their trust. "Multiple excursions were organized," Jollois wrote. "General Caffarelli was eager to seize the occasion to lay hands upon different people who, up until the present, had escaped him. He made out lists, and assigned jobs and destinations at his pleasure. It was a truly contemptible thing to see the officers giving, at their pleasure, destinations to the diverse members of the Institute and the Commission." Jollois was further disillusioned when he saw Monge and Berthollet urging one engineer to accept an especially odious assignment in some distant outpost. "[They] tried to persuade him that he would be doing an operation that would bring him immortality, but it was too easy to see their intentions. On principle, they don't oppose any of the generals." To underline the importance of the effort, Napoleon gave LePère, the thirty-nine-year-old military engineer in charge of the canal survey the heroic (if ridiculous) title "Chief Engineer of All the Works Between the Two Seas."

Most of the recent graduates were dispatched to the desert to help survey the site of the proposed great canal linking the two seas. Devilliers, Jollois, and Dubois-Aymé, at different times, were all assigned to these canal-surveying expeditions. They went, armed with substandard tools salvaged from wreckage or fashioned in Cairo, into a desert infested with human and insect predators that their military guards could do little to repel.

By this point, the desert Bedouins had good reason to attack the French. Napoleon had made it official policy to steal camels from the nomad encampments in order to outfit his new Dromedary Regiment, formed to supplement his cavalry in preparation for a campaign north into the great swath of Levantine turf then known as Syria (now Israel, Lebanon, Syria, and Jordan). Adding insult to robbery, Napoleon envisioned his camel cavalry as a desert force modeled on the nomadic Arabs themselves. The French camel mounted forces were given spears and dressed in gray uniforms, Arabian cloaks, and turbans.

The engineers and surveyors in the desert paid dearly for these military incursions, as Bedouins began shooting at any French they encountered. Even apart from Bedouins hunkered down in the wadis, at the best of times sand filled their instruments. They were also painfully aware that the crudity of their methods raised doubts about the ultimate value of their results. Their leveling telescopes, for example, were marked in decimals (a scale of tens), while their surveying stakes were marked in duodecimals (a scale built on multiples of twelve). They knew they needed to double-check their work, but short food and water supplies drove them back to Cairo.

The military ignored the surveyors' complaints. After one failed excursion in February 1799, Jollois faulted Caffarelli in particular. The returning civilians, he reported, "went to find General Caffarelli. They were welcomed with pretty words. The general needed camels, and he found the means of taking some from them, without being short of politeness. He had promised them some money and never gave them any. He had accorded them an escort and it was taken away from them for some time; he put it out of their power to continue their work."

Such deprivations notwithstanding, the engineers forged into the desert three separate times with the aim of surveying the great canal. They often found themselves literally in the thick of battle, scurrying for cover as skirmishes broke out between their guards and bands of Bedouins or local Arab insurgents. On one trip to Lake Menzalah in the delta region in November and December, the engineers attempted to measure the contour of the lake, while the local Arabs mustered a mini-rebellion against the French, provoking "the most horrible disorder," Jollois wrote. Arab snipers picked off French soldiers one by one in the unsecured area, and the French wanted to surrender the region to insurgents. "The French, intimidated by the revolt of all of the inhabitants of the surrounding country and the insurgent fishermen of the lake, were determined to abandon Lower Egypt. The difficulties that the engineers experienced during this survey were without number. More than once, they were obliged to walk naked in mud and slime up to their knees." When they weren't knee deep in mud and dodging bullets, Jollois added, blowing sand blinded them.

In the desert east of Cairo, Devilliers was part of another unfortunate field crew whose military guards galloped off without warning, leaving them to work unprotected. One Arab guide even led the thirsty French away from a well in order to protect his own tribe's passage into Syria. That expedition ended with a desperate sixteen-hour march back to Cairo, fleeing rumors of marauding Arabs from Suez. On this trip, Devilliers despaired of using his substandard, sand-clogged tools and turned to an ingenious, indigenous measuring system. "The step of a camel is of a perfect regularity, it is a true animal clock," he wrote in his journal, describing how he mapped the desert area known in the Bible as the Wilderness of the Wandering, between Cairo and Suez.

Chief Engineer LePère was fully aware of the possibility for error by his poorly equipped field officers. He ordered his surveyors to cover the ground twice or even more often if they had even a single grain of doubt as to the accuracy of their calculations. He knew that subsequent surveys of the same terrain were extremely unlikely. His engineers rarely, if ever, were able to comply with the order.

The engineers brought numbers out of the desert. The trouble was, they led to a false conclusion.

In late 1800, LePère finally produced for Napoleon an interim report that argued not for directly linking the seas but for connecting the Red Sea via the lakes in delta Egypt to the Nile. LePère's reason for choosing this circuitous route reflected a huge surveying error. He and his engineers had determined that the Red Sea was thirty-three feet higher than the Mediterranean. Thus, the French survey reinforced the faulty seventh-century Persian notion that any canal between the two seas would inundate Egypt with salt water.

Not all the scholars accepted LePère's conclusion. Fourier correctly calculated that such a sea-level disparity could not coexist with the known laws of nature concerning water levels and equilibrium, but he refrained from publicly expressing his disagreement until he was safely back in France. By the time the final canal report was published in 1809, the French had turned their attention to more immediate problems in Europe and Napoleon cared little about a ditch in the desert.

French surveyors finally corrected the mistake with new measurements in 1847. Devilliers, one of the few original expedition engineers still living, was outraged, insisting that he and his colleagues had been right. In the 1860s, another set of engineers, led by the Frenchman Ferdinand de Lesseps, actually began digging what we now call the Suez Canal, more than half a century after the first survey. The world—then and now—might have looked much different had Napoleon's engineers calculated correctly in 1800 and the French succeeded in building a canal as the general was approaching the height of his power.

Napoleon's engineers failed spectacularly in the matter of the canal, but they did find an oddly inscribed, medium-size, pinkish rock that would reveal a great lost swath of human history. The key that eventually unlocked ancient Egyptian writing was discovered in Rosetta on one very hot but otherwise ordinary summer day in late July 1799. The French were digging along a wall, trying to reinforce an old Crusader fort on the west bank of the Nile. In the debris, a French officer of the engineers named Pierre-François Xavier Bouchard noticed an inscribed granite stone. Brushing away dust, he saw carved writing in three scripts, one obviously Greek and one clearly hieroglyphic. Bouchard reported the stone to the general in charge at Rosetta, Jacques Menou, who took it into his own tent, had it cleaned, and arranged for the Greek to be translated. Menou ordered the engineers to search for more fragments of what everyone agreed was a very significant find, but even though soldiers were told such bits would be "worth their weight in diamonds," they found no more pieces.

A few days later, the civilian engineer Michel-Ange Lancret sent a letter to Monge and Berthollet in Cairo, informing them of the find. Menou shipped the stone to the scholars in mid-August. The expedition's Orientalist, Jean-Jacques Marcel, examined it first. He identified the middle script as "demotic"—a popular, comparatively simplified version of the ancient Egyptian language. The scholars were thrilled. Possibly, the fifty-four lines of Greek and thirty-two lines of demotic on the stone would make it possible to translate the third, smaller, fragment of text, fourteen lines of hieroglyphic script at the bottom.

When the French found the stone, it had been 1,500 years since any living human could read a single character of Egyptian hieroglyphic script. Although the ancient civilization lasted for thousands of years, the eradication of Egyptian paganism was abrupt and complete. The Christians who conquered Egypt in the third century ordered the Egyptians to stop using their ancient religious writing. The Copts, newly Christianized Egyptians, submitted without protest and switched to the Greek alphabet. With the ancient culture already in serious decline, it was only a generation or two before no living person could decipher "the writing of the gods," as the hieroglyphic writing was known.

The ancient Greeks first promoted the idea that the symbols represented magical knowledge rather than individual letter sounds, and the medieval Arabs agreed. They deemed the mysterious Egyptian markings to be aspects of *al-keme*—Arabic for "Things Egyptian," from which we derive the English word *alchemy*, a medieval precursor of the modern science of chemistry. Renaissance European scholars taught that the hieroglyphs were magical symbols, and some of Napoleon's scientists flirted with, if not entirely shared, that idea. The mystical notion of "pyramid power" persists to this day.

Finding the stone was a heady moment for the French scholars and soldiers alike. In September, *Le Courrier* reported the finding of "the key" to the hieroglyphs. Nicolas Conté, using the stone as if it were an engraved plate, made prints with the hieroglyphs in black on a white background. (The ink he used for this process turned the stone black.) The Institute sent some of Conté's prints back to Paris. Surprisingly, they evaded the English blockade. News of this fantastic find was the last official interaction between the scholars and their homeland for nearly two more years.

CHAPTER 6

THE DOCTORS

> A long train of loaded camels made their way towards the
> Citadel where everyone was looking for shelter; the khamseen,
> with its whirlwinds of dust, covered all objects with a somber
> veil and gave a livid color to the sun itself: many rich funerals
> crossed the plain, and the cry of professional mourners made
> itself heard from time to time. A Turk, driving a donkey upon
> which lay the cadaver of a French soldier, passed near me. A
> man advanced with long steps with a basket on his head. He
> murmured the funeral song of the Muslims; children's little
> arms and little legs that hung out of the basket showed me that
> the same scythe cut down the rich and the poor, the strong and
> the feeble, at the same time.
>
> —Dubois-Aymé, in *The Description of Egypt*

Syria, Spring and Summer 1799

La peste, the French called it.

Field doctors rarely uttered the name; they thought it
was too demoralizing. To Anglophones, it sounds rather benign: pests
are minor annoyances—flies, gnats, fleas. *La peste*, though, is etymo-
logically related to our longer word, the more ominous *pestilence*, one of
the Four Horsemen of the Apocalypse. During the French occupation,
the bubonic plague epidemic in Egypt was a killer of biblical stature,
a germ that caused men to die hideously, rotting from the inside out,
sometimes within forty-eight hours.

The ancient Egyptians knew about the plague. They believed it emanated from the goddess Sekhmet, a lion-headed female deity who presided over war, violent storms, and pestilence. When appeased, Sekhmet healed. When offended, she destroyed. A group of black stone Sekhmet statues lay buried in the sand at Karnak, an ancient site along the Nile that the French scholars would explore. Years before anyone alive understood the goddess's significance, the French *savants* wondered at the identical, stone feline faces, their dull surfaces impervious to human suffering.

Of all the foes the French faced in Egypt—Mameluke warriors, the English navy, Ottoman armies, Arabs in revolt—the most terrifying enemy was invisible until too late. No doctor knew how to prevent or cure it. Scores died trying.

Egypt was a sickly place for the French even without the plague. Mundane problems included camel bites (sometimes rabid, and common during the beasts' "love season"), scorpion bites, snake bites, internal leech bites (from drinking infested water), bronchial infections, worms of all types, venereal diseases, and fevers. The other virulent diseases, much less devastating than the plague, included dysentery, which killed nearly 2,500 men, and ophthalmia, an extremely painful but not deadly eye infection that glued eyelids shut with pus and rendered men temporarily blind. Ophthalmia afflicted almost every French man in Egypt at least once.

Plague, however, was catastrophic and terrifying because it was almost certain to kill. Ten out of twelve of the French unlucky enough to catch it died. The signs were easy enough to recognize. First a man felt nauseous. He had a bad taste in his mouth, vomited, then was hit with a blinding headache and a fever. A few hours or days into the waxing delirium, unmistakable signs of *la peste*—the bluish lumps called buboes—erupted in the armpits and groin. Eventually, these lumps exploded, oozing pus and giving off a stench of decay, by which point most of the stricken were, mercifully, barely conscious.

Napoleon was not oblivious to the dire health situation in Egypt. His first order of business in Cairo was to establish new French hospitals. Once they were operating, he ordered musicians to play martial music outside them every day at noon, so the confined patients "would be inspired with gayness and be reminded of the beautiful moments of past campaigns."

He did not appreciate until too late that disease, not war, would kill a third of his army and help doom his Oriental project. Of the 50,000 men who went to Egypt with Napoleon, possibly 10,000 perished of disease.

The Institute's medical members routinely delivered reports on the various ailments indigenous to Egypt. The scholars, however, usually learned about specific health perils through personal experience. Most were, at some point or another, severely ill in Egypt. At the very least, they were temporarily blinded and suffered from fevers, rashes, and diarrhea. Only ten died of plague, a much smaller percentage than among the soldiers, possibly due to the fact that they were shielded from the contagion by reason of their privileged living quarters.

By the end of the occupation, most of the French had seen plague victims as cadavers carted through the streets. The chief eyewitnesses were the doctors and surgeons, who dealt with the disease firsthand and delivered regular reports to the Institute. Their observations offer a harrowing picture of the daily lives of the French in Egypt, menaced on all sides by insects and diseases unlike any known in France, with remedies equally mysterious, if they existed at all.

The expedition's two chief military health officers were prominent, active members of the Institute. The army's physician-in-chief was the aristocratic, quick-witted René Desgenettes, best known for overseeing military hospitals in the regions Napoleon conquered. The chief surgeon, Dominique-Jean Larrey, a son of the middle class, led the military surgical units and is remembered as the inventor of the rudimentary ambulance.

The two medical men—close contemporaries in age, thirty-seven and thirty-two, respectively—were a courteous working team, but they didn't much care for each other. "They were two natures totally unlike, and their origins, their education, their tastes, and their characters distinguished them profoundly," wrote Desgenettes's biographer. Larrey was prominent not by birth but by his own works, and he had a certain unrefined quality that might be expected in a surgeon in the era before anesthesia. He was disciplined but lacked the subtlety that a socially prominent family and expensive education had given his counterpart.

Larrey studied surgery as a college student. Twenty-three when the Revolution erupted, he had already made a name for himself as intelligent and mature. He taught and practiced in Paris during some of the most turbulent years of the Terror, and was appointed surgeon-major of the Army of the Rhine in 1792.

Larrey's contribution to eighteenth-century medicine was the "flying ambulance," a horse-drawn cart that trundled on and off battlefields with first aid for the wounded. Before Larrey's time, wounded men were left on the field, sometimes for days, before getting help. In Egypt, he altered his flying ambulance to fit camels, yet another use for the versatile beasts the French were stealing from the Bedouins.

Desgenettes was cultured and astute, born in 1762 into an old, aristocratic family. His father was a lawyer and his mother a member of the lesser nobility. He was, in fact, *Baron* Nicolas-René-Dufriche des Genettes until the Revolution. After one of his uncles was guillotined in Paris during the Terror, he hid for months in rural Normandy. Returning to Paris, he dropped his title and changed the spelling of his name to the more democratic single word. He never did shed, however, the identifying characteristics of his class. "He owed to the milieu in which he was born and in which he had spent his youth, the ease of manners, the studied elegance of language and style, the perfect courtesy, and even a sort of elegant and sloppy cynicism that characterized the doctors and the socialites of the end of the 18th century," wrote his biographer.

Desgenettes joined the army mainly to keep his head below the revolutionary scythe. His confident skepticism and caustic sarcasm put him at odds with Napoleon on many occasions, and their public tiffs became part of the Institute's record. An able doctor and administrator, Desgenettes also loved words and speaking. In addition to his medical duties, at Cairo he presided over the publication of the Institute's journal, *La Décade Égyptienne.*

Between them, Larrey and Desgenettes were charged with overseeing the health of the entire French army in Egypt. As the situation deteriorated, Napoleon put them in charge of deciding which of the numerous officers begging to be sent home for health reasons was actually sick. In this capacity, both men proved above bribery. As Institute members, the two doctors and their staffs also studied Egyptian diseases and indigenous medicine. Desgenettes ordered military doctors to investigate and record traditional Egyptian medicine, and to weigh the effectiveness of native means of curing indigenous illnesses of which the French had little or no knowledge.

In their report on Egyptian medicine in *The Description of Egypt,* the authors noted that the Egyptians had a few recipes for medicines dating to the era of the great Arab physicians in the Middle Ages, from whose legacy European doctors had learned much. However, the French found that modern Egyptians often went without medicine, and were generally rather fatalistic about illness, which the authors attributed to their strong religious beliefs. "Most of the Muslims are persuaded that everything is predestined and do not really believe in the curative properties of medicine and medical practices," the *savants* wrote. "So, as long as they have followed the precepts of cleanliness and sobriety, if they get sick they think it is sent from God, but bear their illness with bravery and without complaint, and if they finally have recourse to medicine, it is often too late for it to be of any use. This fatalistic attitude has caused medical science if not to diminish, at least to stop its progress, and this in the country where this science was born."

The French did catalogue a few folk remedies, but they showed rather more interest in Egyptian recreational drugs, which they divided into three categories: opiates, "to procure real or ideal pleasure"; drugs put in the bath that were supposed to enhance sexual pleasure or ability (the writers noted that the Egyptians "seemed to have insatiable appetites for voluptuousness and orgasm"); and drugs used as cosmetics, to whiten the skin or dye the hair.

The French attitude toward Egyptian folk medicine was supercilious. It was "blind and brutal, performed by ignorant, presumptuous barbers," wrote one medic. Of course by modern standards eighteenth-century French medicine qualifies as "blind and brutal." Besides having no clue as to how the plague was transmitted (via fleas, carried in rats, clothing, or bedding), they clung to the medieval notion that diseases were caused by the "suppression of transpiration"—the passage of water in and out of the body. Larrey believed that the blinding eye infection ophthalmia was caused by abrupt temperature changes—from daytime desert heat to nighttime cold, for example—which stopped up a good sweat. Desgenettes, too, shared the idea that a good sweat cured the direst disease.

Their "cures" were ghastly experiences, even if they occasionally seemed to work with patients who happened to recover thanks to luck or hardiness. To effect "transpiration," the doctors administered purges or emetics of various strengths to even the most seriously ill and wounded men. Larrey routinely applied leeches and scarification to the temples of men afflicted with agonizing eye conditions. Descriptions of his field surgery, including amputations at the hip and shoulder without anesthetics, are almost unbearable to contemplate.

Almost as soon as they landed, the French saw their first plague victims among Egyptians. The contagion nipped at the poorer sections of Cairo that first fall, but did not immediately spread in the city. *La peste* first afflicted the French in Alexandria at the beginning of December. Within two weeks, thirty were dead, and less than a month later the plague had claimed another 130 more French lives.

Besides applying olive oil to the skin and administering purgatives, the doctors had two other means of combating the plague—vinegar and fire. They burned the clothes and bedding of the sick, which coincidentally did kill the fleas, and disinfected rooms and tools with vinegar, which did nothing but leave a sour smell. Ignorant of the fact that rats and their fleas were the primary plague carriers, men remained vulnerable everywhere in the field, especially when they recycled the clothing and blankets of the dead.

Desgenettes at least had the right instincts. He urged soldiers to burn the clothing or linen of plague victims, not to reuse them. But the soldiers ignored his advice, and the officers, wanting to economize, hesitated to approve the destruction of perfectly good uniforms and bed linen. Napoleon tried to intervene. "I have come here to fix the attention and transfer the interest of Europe to the center of the ancient world, and not to pile up wealth," he pleaded in one order backing Desgenettes. After the examples of plunder in Italy and Malta, the soldiers laughed at that order and continued hoarding dead men's things.

Clothes were not disposable, but Egyptian women apparently were. After men began dying of plague at Alexandria, the French instituted a draconian precaution at Cairo: all prostitutes found having relations with a Frenchman were ordered enclosed in a bag and thrown into the water. The massacre of street whores was intended, to be sure, as a prophylactic measure. While the French didn't really understand how plague was contracted, they knew it had something to do with where, in what, or with whom men slept. Keeping men in their own clothes and their own bedding probably did save some from getting sick. Egypt's prostitutes were not a source of plague—plague-infected prostitutes by definition would be too sick to be servicing French troops—but their beds might harbor disease-carrying fleas.

The plague killed such a large percentage of those who contracted it that the French soldiers came to fear it in an almost superstitious way. The longer they stayed in Egypt, the more theories they devised to explain it. They thought it had a particular season and climate (although it really had none, depending, as it did, on the rat population). They thought it could be warded off or made milder if a man simply put his mind to ignoring it. Many of the scholars themselves adhered to the latter notion, believing that the disease was curable with exercise and good spirits, and that it killed only those who feared it most.

Doctors who saw the disease up close could be at least as terrified of *la peste* as the rank and file. A surgeon named Boyer refused to treat plague victims at the hospital in Alexandria. Napoleon, hearing of this, ordered him arrested, dressed in women's clothes, placed on horseback and paraded through town with a sign on his back that read: "Unworthy of being a French citizen, he is afraid of dying." In his memoir Desgenettes noted that this punishment infuriated one French major's wife. "She was highly incensed over the fact that the wearing of women's clothes had been ordered to symbolize cowardice."

The elegant Desgenettes proved himself among the braver doctors, but he had no better idea as to the real cause or cure of plague than anyone else. He believed the best plague cure was a head-to-toe coating of olive oil and a good sweat. He warned attendants applying the oil to wear protective oilskin suits and wooden shoes, and to rub their own bodies with a coating of olive oil before approaching the patient. In hundred-degree-plus desert heat, the treatment ensured epic perspiration in both patient and nurse. It probably didn't act as a flea repellent.

L*a peste* was still just lurking in the shadows when Napoleon mustered 13,000 men for a preemptive strike into what was known as the Holy Land—a much vaster territory than what we now associate with that term, comprising Israel, Syria, and Lebanon. He marched this army out of Cairo and toward Syria in February, 1799. The British and Turks had by this point come to an uneasy truce—made allies by their common enemy. The British were hovering along the coast, leaving, for the moment, the bloody work of land combat to the Ottoman army, which was marching south from Turkey to meet the French. The British were biding their time in the water, monitoring the death throes, as they believed, of the French campaign, standing by to assist the Turks from sea if necessary. Conventional wisdom held that the French army would eventually have to admit defeat, not attempt greater conquest. At this point, and for many more months, the British saw no reason to assign more troops to engage the French on land.

From the British point of view, Napoleon's decision to invade the large territory also known as Syria was even more ill-considered than the initial invasion of Egypt. The general seemed to have lost his senses, as he marched his men through the desert, knowing full well that his supply line was blown up and all reasonable hope of reinforcements from France was gone. Napoleon apparently had decided he had nothing to lose and everything to gain by forging deeper into Ottoman territory by land.

The general was relying on the commonly held belief that the Ottoman Empire was crumbling. The Ottomans, of course, disagreed with this dire analysis, although the Sublime Porte, as the seat of government was known, was aware of the empire's flaws. Sultan Selim III, a reformer, and known to the Turks as "The Inspired," was enamored of some Western ideas. An influential group in Istanbul admired the French Revolution as a secular experiment that might hold lessons for the future of their realm. The French and the Ottomans were long-time allies. They had made their first alliance in 1536, and the revolutionary French had sent military trainers and technicians to the Ottomans. They had given the Turks a printer and planted a "tree of liberty" in Istanbul to install a symbol of their new values in the heart of the feudal empire. Most members of the Ottoman power structure, however, were conservative (Selim III paid with his life for his reformist tendencies) and had little use for equality, liberty, or brotherhood. In 1792, as the French were chopping off Louis XVI's head, the sultan's own privy secretary, Ahmed Effendi, wrote, "May God cause the upheaval in France to spread like syphilis to the enemies of the Empire, hurl them into prolonged conflict with one another and thus accomplish results beneficial to the Empire, amen."

Disregarding the Turks, Napoleon was motivated by one of France's chief objectives in Egypt: "He will chase the English from all their possessions in the Orient." Taking Syria would be one more step in the direction of India, the big Asian prize in the ongoing colonial competition with the British. Before leaving for Syria, he sent a message to an Indian leader allied with the French, the Sultan of Mysore, urging collusion now that the French were in Asia. The British prevented that possibility by seizing the rebellious Indian's territory and killing him. Napoleon lived to regret the Syrian gambit at least as much as he regretted leaving ships in Abukir Bay, but he never acknowledged his errors.

Eleven members of the civilian Commission accompanied Napoleon into Syria, including Monge, Berthollet, and Savigny. After "Mongeberthollet's" brush with death at the hands of the Mamelukes in the Nile, Napoleon refused to be separated from his favorite pair of scholars and took them with him everywhere—an indulgence they must occasionally have regretted. During the return trip from Syria, Monge became gravely ill, possibly with the plague, but recovered, a fact he later attributed to Berthollet's devoted care in a field hospital.

Marching north through the barely mapped Sinai Desert, the French soldiers were again in a state of abject, burning rage. For them, the glory of winning a geographical chess game against the English was hardly worth the misery of the desert march. They met no English on the way, and the idea that they might be severing a British land route to India was not reward enough when they were dying of thirst. Men punctured the water casks with their bayonets, provoking general disorder. "They insulted those whom they saw on horseback," wrote Napoleon's secretary, Bourrienne. "They indulged in threatening language against the Republic and the *savants*, whom they regarded as the authors of the expedition."

The French clashed with the Turks at a famous battle at Al-Arish, on the Mediterranean coast in the Sinai Peninsula. Larrey's camel-drawn ambulances went onto the field for the first time there. Less inclined to openly question his general than the skeptical Desgenettes, the chief surgeon had so far kept his unhappiness in Egypt a secret. "If you only knew, my dearest Laville, the extent of our privations and miseries and what I have suffered during this unhappy expedition," he wrote his wife shortly after first arriving at Cairo. "But fate is not done with us yet! I am one of those who are shackled by unbreakable chains to the chariot of our modern Alexander."

Fate held much worse in store for the miserable surgeon. As Larrey attended the sick and wounded among the Turks after the first battle, he found wounded men inside the enemy garrison with worms in their flesh, and buboes in their groins. The oozing sores and dying, feverish men "showed me clearly that plague was present amongst the troops of the enemy," he wrote in his journal. The French army was marching directly toward a rendezvous with *la peste*, and Larrey could do nothing but bring out the olive oil and vinegar.

At Jaffa, several days' march up the coast from Al-Arish, the French besieged another Turkish garrison for four days. Once inside the city, they ran amok, releasing their rage at their own situation by raping and pillaging. The young engineer and Institute member Étienne-Louis Malus witnessed the orgy of violence. He later published a vivid account in a brief memoir of the Syrian campaign, *L'Agenda de Malus*. "The tumult of the carnage, the broken doors, the homes shaken by the noise of the fire and of arms, the screaming of the women, the father and infant dumped one on top of the other, the raped daughter on the cadaver of her mother, the smoke of the dead scorched by their clothing, the smell of blood, the moaning of the wounded, the yells of the victors disputing their plundering of an expiring prey, furious soldiers responding to the cries of the wounded by cries of rage and repeated blows, finally of men satisfied of blood and gold, falling of weariness on the heaps of cadavers: there is the spectacle offered to the unhappy town that night."

Four thousand Turks surrendered at Jaffa, on condition that they be allowed to live. Without food, or manpower to police them, Napoleon made a decision there that would become one of the darkest elements of his legacy. He ordered the defenseless, unarmed Turkish prisoners taken to the seashore and killed. To save powder and bullets, the general ordered his men to massacre them with their bayonets, which they did, chasing those who tried to flee into the surf and stabbing them until the sand was stained red. "The atrocious scene makes me yet shudder when I think of it," wrote Napoleon's own secretary, Bourrienne, describing how the French soldiers used the Turkish sign of truce to lure the escaping swimmers off the rocks only to attack them again with their bayonets. "All that can be imagined of this day of blood, would fall short of the reality." Travelers claimed Jaffa still stank of rotting corpses two years later.

As the ancient Egyptians had divined, war was linked with storms and pestilence. Shortly after this shameful massacre, *la peste* visited the French ranks with a vengeance.

Malus, twenty-three years old, explicitly linked the French brutality at Jaffa and the plague. "The frenzied pillage delivered a miasma. It was contained in the clothes they had greedily taken. The mortal effect was rapid. The illness appeared on the battlefield, in buboes and carbuncles. The horrifying cry 'It's the plague!' spread through the army and struck terror in the most courageous and invincible."

The young scientist experienced Jaffa as both an army engineer in charge of a makeshift hospital and then as a patient in it. Like Desgenettes, Malus, a son of one of Louis XVI's financial advisors, had upperclass origins. He joined the army as his only option after banishment from school for his blood ties to the nobility but later was allowed to study at the École Polytechnique when his military superiors noticed his special intellectual abilities. Monge took a personal interest in the bright young man, giving him three months of private lessons. Malus was an unusually scholarly and literary engineer. While in Egypt, he began laying out a theory of the polarization of light that later earned him lasting scientific recognition back in Paris.

In his memoir of the Syrian campaign, Malus includes a gruesome account of the Jaffa plague hospital and his own illness. His first task was to transform a Jaffa convent into a hospital. It was also his duty to create a means of disposing of the growing heap of corpses. "Immediately after the departure of the army, we barricaded the doors, we repaired the holes, and we occupied ourselves to bury the cadavers that encumbered the streets and houses." Through contact with the bodies of plague victims, he became infected himself ten days after starting to work among the dying and the dead. "I had assiduously gone and passed the morning in the infected odor of this cesspool, the least corner of which was filled with the sick," when he noticed the telltale symptoms—fever and headache—of *la peste*. "Dysentery continued and bit by bit all the symptoms of the plague declared themselves."

While Malus sank into delirium with other unlucky French soldiers at Jaffa, Napoleon and the able-bodied marched on up the Coast. By March 15, they reached an old Crusader fort called St. Jean d'Acre. The fort was named after the Knights of the Order of Saint John, who retreated to it after Jerusalem fell to the Saracens in 1244. The fort figured in the final defeat of the Christians in the Holy Land, and all sorts of dark legends attached to the site. One of its towers was called the Fly Tower because it was supposedly built on the site of a temple dedicated to Beelzebub, Satan's Lord of Flies. Another of its towers, known as the Tower Maudite (*maudit* being the French for "accursed"), was said to have been built with the thirty pieces of silver for which Judas sold Jesus Christ. One of the more gruesome episodes in the bloody history of the Crusades occurred at the nearby Acre convent. A group of nuns, seeing the Muslims about to take the city from the Christians, cut off their noses in order to make themselves too ugly to be raped. Thus disfigured, they escaped being sexually violated and were simply murdered, dying with their virginities intact.

When the French arrived at this blighted spot, it was commanded by a notorious Turkish pasha known locally as El Djezzar (the Butcher) and his band of Turkish soldiers. The Butcher had earned his name by performing the usual atrocities—tying prisoners in pairs, sacking them up, and throwing them in the sea; personally chopping off heads—but also for his unusually creative, diabolical imagination. Not long before the French arrived, he had squeezed living Russian Christian prisoners into the crevices of the walls he was reinforcing around the fort, plastering them in so that only their heads protruded from the mortar, thus creating a living wall of slow death. Some of their skulls still decorated the perimeter when the French arrived. In his less inspired moments, the Butcher was known for flaying his discarded wives, chopping off the hands, ears, and noses of his companions, all on the slightest pretext.

Larrey worried more about the plague than about the Butcher. As the French prepared to attack the fort, he sent an order to the army surgeons in the ranks. "Citizens, I beg you to report, every fifth day, the number of sick in your respective units, the character of the prevalent disease." He warned them that the "prevalent disease" was "contagious at a certain stage" and advised soldiers never to use Turkish blankets or to sleep in holes in the sand.

Faced with growing terror of the disease among the ranks, Napoleon, with Desgenettes in agreement, officially denied the epidemic was plague. They agreed to call the sickness among them a mere "fever with buboes." Napoleon shared the scholars' belief that fear of the disease actually made a man more likely to catch it. Desgenettes later wrote that he was never convinced of the relationship between a person's fear of the disease and his susceptibility to it, but to be on the safe side he agreed never to speak of *la peste*. "I thought that I should, in this circumstance, treat the entire army as a patient to whom it is almost always useless and often dangerous to enlighten about his illness when it is very critical," he later wrote. Larrey, on the other hand, was diagnosing it openly.

By the end of March, the French were hunkered down beneath the impregnable gray rock walls of the fort, and *la peste* stalked them openly. They inhabited their own valley of the shadow of death. At night, the camp resembled a Hieronymous Bosch painting. Packs of jackals slunk in, and their cries, added to the howling of stray dogs, the braying of donkeys, and the moans of the sick, supplied a cacophonous soundtrack to the unfolding nightmare. To complete the surreal picture, troupes of costumed Egyptian jugglers and fire-eaters arrived nightly to entertain the troops for money.

Just off the coast of Acre, an English squadron led by Sir William Sidney Smith monitored the French disaster, while supplying the Butcher with arms and food. A French royalist engineering officer named Phelippeaux, who had defected to the English after the Revolution, assisted Sir Sidney Smith. Phelippeaux harbored a personal distaste for Napoleon, with whom he had attended military school. Smith and his men held their moral scruples tightly in check and watched, wide-eyed, as the atrocities proceeded. "The Turks brought in above sixty heads," Smith wrote after one battle with the French, referring without further explication to the Turkish practice of mutilating the bodies of the dead soldiers and keeping their heads as trophies.

A week into the Siege of Acre, Napoleon learned that the British had captured a small French convoy carrying heavy artillery up the coast to his aid. The heavy artillery was crucial to bringing down the walls of forts like Acre. Conté's genius might even have prevented the disaster, had Napoleon listened to him. Before the army left Cairo, Conté had suggested loading the artillery on wagons and driving them on temporary roads created by camels dragging specially designed, wide-wheeled carts. Napoleon wasn't interested in the land-route experiment.

Without heavy cannon or other fresh supplies, the French were reduced to flinging themselves at the fort's rock walls in human waves. Those who managed to scale the first wall found a wide moat on the other side, and the Turks easily picked off those intrepid French who dared try to swim across. Field doctors, already stretched treating the plague-sick, now began receiving hundreds of men hideously wounded at close quarters. From a field hospital set up in the Butcher's stables, Larrey went to work, dressing wounds and amputating ruined arms and legs. He was amazed at how quickly bluebottle flies gathered on the open wounds, laying maggots that hatched within twenty-four hours. Field doctors a century later, in World War I, noticed the maggots might actually speed the healing, if only patients could bear the itching.

General Caffarelli, leader of the engineering corps, was among those mortally wounded at Acre. He was hit by a shell in the arm, and Larrey amputated it at the shoulder. Caffarelli lingered for eighteen days before dying of an infection and fever. He went like the best secular hero, insisting during his final hours that friends at his bedside, among them Monge and Berthollet, read aloud to him the essays of Montesquieu and discuss political economy.

Denon later eulogized Caffarelli in his book. "He mingled the love of humanity with the ardor of bold undertakings, watching incessantly over the welfare and preservation of his fellow creatures; in him every being possessed of intellect or of feeling lost a father and friend; in making my drawings, I have often thought of the pleasure I should have in showing them to him." He was even so well regarded in Egypt that the Arab historian Jabarti wrote of his courage and valor.

Napoleon took Caffarelli's death hard. The older man had been a friend and trusted advisor. He ordered Caffarelli's heart mummified and eventually took it back to France with him.

Being a general, Caffarelli had the privilege of dying on a bed. Most of the sick and maimed died on the stable's floor, and without medicine. The expedition's chief pharmacist, Claude Royer, had failed to deliver drugs and linens to Syria as ordered. Rather than loading supplies for the sick at Cairo, he heaped his camels with liquor, coffee, and other luxury goods, and sold them en route, making a handsome profit for himself. Napoleon ordered him shot for corruption, but the medical corps pleaded his case, and Royer was freed. He later defected to the British side in Cairo, and made himself very popular with his new comrades by describing how he'd poisoned plague victims on Napoleon's orders.

The plague was thinning the French ranks more efficiently than the efforts of the English and Turks combined. Larrey and Desgenettes tried to control the spreading contagion at Acre by shipping plague victims back down the coast to the hospital at Jaffa. There, soldiers were dying of plague at the rate of thirty per day, with new patients arriving daily to replace each one who died. For six weeks, the sick were delivered by the cartload from the Acre front to Jaffa, where they "nourished the warehouse of the dying," Malus wrote. Many of the infected died within forty-eight hours of their arrival.

To lift the morale of his terrified men, Napoleon paid a visit to the Jaffa hospital. In a famous moment, later immortalized by the French painter David, he personally lifted a dying plague victim whose clothes, according to one horrified observer, "were covered in lather and disgusting evacuations of an abscessed bubo." In the painting, David drew Desgenettes keeping a safe distance from the dying man. Napoleon never got sick.

Desgenettes also tried his hand at the public relations cure. At the Acre field hospital, he stood before doctors and soldiers and pierced himself in the groin and armpit with a scalpel dipped in bubo pus, to make the public point that a man might recover from the disease. He promptly washed himself with soap and water, though, and never did contract plague.

Delirious in his sickbed at Jaffa, Malus was beyond the reach of any propaganda gestures. He was barely alive a week after the telltale blue lumps had emerged in his groin and was sure he would perish soon. He made arrangements for imminent death, sent his personal things away to friends and watched listlessly as visitors and servants caught the plague and died around him. "I was alone, without strength, without help, without friends. I was so exhausted by the dysentery and the continual suppurations, that my head was extraordinarily enfeebled; the fever, which intensified at night, often gave me le transport and agitated me cruelly. Two sapeurs who tried to take care of me perished one after the other."

After three weeks, help arrived by sea. Malus was loaded with other patients onto a boat sailing for Egypt. The boat's captain already had the plague and died soon after they weighed anchor, yet Malus began to recover on the ship (he credited the "sea air"). By the time he reached Egypt he was able to eat again. On shore, he was clapped into a quarantine hospital that was without doctors and "stacked with plague victims," he wrote. Sick and well shared the same rooms. The uninfected "got it right away" and "perished to the last," Malus wrote.

"It was rare that one left this infernal prison when one had the misfortune to enter here," he wrote. "I heard them often die of rage, demanding water of the deaf who pretended not to hear them or who responded: it's not worth the trouble. The avid gravediggers stripped the moribund before their last sigh. These barbarous agents of the sanitary commission stood in for doctors. Hardly had their victim ceased to move, then he transported him to the other bank where he abandoned him to the dogs and birds of prey. Often he covered them with a little sand, but the wind soon uncovered the cadaver and this refuse looked like a hideous spectacle of a battlefield."

Malus survived the plague, but doctors in France believed it left a lasting vulnerability in his body. He returned to Paris and was made a professor, but died young—in his mid-thirties—from a mysterious infectious fever that also killed his wife.

The siege of Acre lasted six weeks. French soldiers scaled the fortress walls sixteen times before Napoleon finally accepted the futility of it and ordered a retreat to Cairo. On May 21, the soldiers, including the sick and wounded who could walk, started marching south toward Egypt. The dying—dozens or hundreds depending on whose account one reads—were left behind, subject to an infamous act of euthanasia. Napoleon, arguing that bringing them along in retreat would subject the healthy to contagion, while leaving them behind alive would abandon them to torture by the Turks, ordered Desgenettes to administer a killing dose of laudanum to the doomed men.

Desgenettes was appalled. In his memoir, he wrote: "General Bonaparte had me called, early in the morning, to his tent, where he was alone with his Chief of Headquarters. After a short preamble on our sanitary situation, he said to me: 'In your place, I would at the same time finish the plague victims' suffering and end the dangers with which they threaten us by giving them opium.' I simply answered: 'My own task is to conserve.' 'I am not looking,' he continued, 'to conquer your reluctances, but I believe that I will find people who will better appreciate my intentions.'"

The people Napoleon found included the treacherous pharmacist Royer, who later told the English that he administered the drug himself. The English soldiers who surged into Jaffa after the French retreat claimed that more than five hundred French patients had been euthanized. The subsequent reporting of this event in newspapers across Europe permanently tainted Napoleon's reputation. The French officially denied it, but eyewitnesses confirmed the deed, although they put the number of victims much lower. Desgenettes believed that at Jaffa twenty-five or thirty of the plague victims had been given a strong dose of laudanum but that not all died. "Some rejected it by vomiting, survived and later told all that had happened," he wrote.

Napoleon himself categorically denied having ordered poisoning, and blamed the story on "a misunderstanding." Desgenettes, though, never forgot the request, and his ill will toward Napoleon soon erupted in a very public argument before the Institute of Egypt at Cairo.

The French left Acre thoroughly demoralized. Twelve hundred men had died there—seven hundred of plague, five hundred of wounds—and another 1,100 were alive but wounded. Sir William Sidney Smith, who watched the French withdrawal from Acre, described it in a letter home. "The utmost disorder has been manifested in the retreat, and the whole track between Acre and Gaza is strewed with the dead bodies of those who have sunk under fatigue or the effect of slight wounds."

Inland, the summer heat was again intense, and water scarce. Marauding French soldiers carried torches and vengefully set fire to everything in their path—villages, farms, crops. The noble army of liberation became a desperate army of subsistence. Men threw the wounded from their litters, leaving them to moan and die in scorched fields. Napoleon finally jettisoned his large-caliber guns and ordered every horse, camel, and donkey to be given to the sick and wounded. He personally refused, however, to put a plague victim on his own horse.

The final blast of divine retribution on the French came in the form of the *khamsin*. Sudden, powerful, and deadly sandstorms, they first appear as a blood-red stain in the distant sky, harbinger of waves of suffocating dust. Camels and Arabs, seeing the swelling tint on the horizon, knew to stop and take cover—humans under blankets and camels crouching on the ground with their noses aimed away from the encroaching storm. The French did not react until it was too late, then choked and fainted in the blinding, suffocating walls of dust.

Larrey was "a mournful witness" to all of it, from the sacking of Jaffa to the plague at Acre and the final French retreat, with the shameful abandonment of the sick. Larrey thought he would die during the *khamsin*. "After a few minutes of torture I collapsed in a faint without hope of reaching Salahieh [a fort at the frontier of Egypt]. Many of our animals were suffocated, especially the horses, and the whole army suffered a great deal. This day saw the death of many of the convalescents from plague who were following us."

Plague and *khamsin* didn't complete the torment. When the French finally found fresh water again, it was inhabited by a tiny species of leech, each about the size of a horse hair. Thirsty men guzzled the leeches with the water; subsequently the creatures attached themselves to their throats, where they swelled to the size of normal leeches.

Larrey later read his fellow scientists a report on the effects of the ingestion of leeches. "Our thirsty soldiers threw themselves down at the water's edge and drank freely, without suspecting any new enemy. Many of them felt at once a prickling sensation from the leeches they had swallowed. The immediate effect of the punctures was to produce a painful tingling in the throat and a desire to vomit. Gargles of vinegar and salty water generally sufficed to detach the leeches but at times it was necessary to use forceps."

The scholars back in Cairo remained sanguine about the plague, and blissfully ignorant of leeches and killer sandstorms. When *la peste* first appeared in Alexandria, they were unbothered. In his journal, Jollois said the French greeted news of the disease with much black humor, if not wholesale callousness. "It is said that the plague was in

Alexandria, that it has manifested itself in the hospital of the marines, that all the Frenchmen were camped outside of the city. The letters coming from this [part of the] country were soaked in vinegar. This news threw a bit of sadness among the French, but it did not last long. As early as the next day, they laughed about it, as they usually laugh about everything."

Geoffroy Saint-Hilaire, whose brother, Marc-Antoine, was an officer in the army, wrote a letter to his father in June 1799, reporting that Marc-Antoine had survived a mild "glandular fever" from which men easily recovered. In fact, Marc-Antoine had been one of the lucky few to have recovered from plague. Geoffroy Saint-Hilaire wrote their father that his brother now joked "like the rest of the army about this blight; they call it the *maladie de la glande* and no longer fear it."

Black humor permeated the scholarly corps. According to Desgenettes, Dutertre, the artist who had the job of sketching the scholars and generals, claimed to be more distressed at having been robbed of his subjects than by the horrible events in Syria. "Dutertre would ask, 'How is so and so?'" Desgenettes recalled. "'He's dead.' 'The devil! Too bad! I haven't done him yet.' And of another, 'He's dead, too.' 'That's all right, I've got him.'"

The aristocratic doctor and the general publicly clashed during the first meeting of the Institute after the army's return from Syria. According to the engineer Pierre-Dominique Martin, a member of the Commission, Desgenettes, still furious about Jaffa, reacted to Napoleon's brusque, condescending attempt to dismiss a discussion about chemical processes among the members. "Bonaparte, impatient at some discussion he found trivial, ended it by saying: 'I can see that you are all holding each others' hands. Chemistry is the kitchen of medicine and medicine, the science of assassins.' Des Genettes looked at him fixedly and replied, 'And how would you define for us what conquerors are?'"

A few minutes later, the general and doctor almost came to blows. The scholars were discussing ways to keep the army apprised of the plague. Napoleon still wanted to conceal the sickness's real identity. Desgenettes insisted it was time to end the policy of prevention-by-subterfuge, as a contagion in urban Cairo could be disastrous. Napoleon disagreed. Desgenettes then brought up the matter of the poisoning of patients at Jaffa, and with such fury that members of the Institute feared for his life. Napoleon, though, backed down, albeit "pale with rage."

As much as he disagreed with the general, Desgenettes never left Napoleon's service, although he tried to get sent home immediately after their public clash, on the grounds of poor health. Napoleon refused. The general eventually rewarded him with a barony, which Desgenettes accepted with grace, surely aware of the irony in being retitled after having cast off his noble birthright to save his own life during the Revolution.

CHAPTER 7

THE MATHEMATICIAN

*My friends, if we leave for France, we didn't know anything
about it until midday today.*
—Monge's last words to the scholars
before leaving them in Egypt

Cairo, Summer and Fall 1799

On the night of August 15, 1799, the scholars of the Institute of Egypt were not enjoying another scientific debate among the orange blossoms in their "flaming core of reason." On the contrary, they were arguing about rumors that their modern Alexander was about to cut and run, leaving his scholars and soldiers behind. Napoleon had been back from Acre for almost two months, and had celebrated his "victory" in the Holy Land with a public fête. The pretense had fooled no one, least of all the sickly, shoeless army, which was approaching mutiny.

Everyone knew the situation of the French in Egypt was insecure. But leave? Napoleon? Unthinkable.

In early August, Sir Sidney Smith happily shared with the French some European newspapers from May and June: France was at war with Austria again, and its forces had been kicked out of Germany and Italy, reversing Napoleon's finest military achievements. The Directory was tottering, and a complete overthrow of the Republic seemed imminent. Reading this news just after the disaster at Acre, Napoleon decided he couldn't waste another minute chasing Alexander's ephemeral legacy to the East. He needed only a moonless night and the temporary absence of British ships off a section of the Egyptian coast to sneak away and get himself back to Paris, where, he reasoned to his comrades and close aides, he was more immediately needed. Luck was on his side. The English—who would have loved to sail home with the little French general as a trophy—were watching another part of the coast and missed their chance again. The general sailed away ignominiously with a small party in a ship with all lights extinguished.

Napoleon invited a lucky trio from the Institute—Berthollet, Monge, and Denon—to leave Egypt with him. The three men guarded the secret as best they could, but the small, gossipy community of scientists at the Institute was growing suspicious. A day earlier, at the Institute's weekly meeting, the notion would have seemed implausible. At that time, Napoleon had ordered two groups of scholars to explore Upper Egypt, giving them detailed instructions and behaving in all respects like a man committed to the enterprise. Later that afternoon, though, he'd sent a case of sugar, wine, coffee, and spirits to a waiting French ship under the code label *"Pour Monsieur Smith."* Thus assured of luxury provisions for himself on the way home, he hosted a lavish dinner for members of the Divan, inviting Monge and Berthollet to join them.

Back at the Institute, the civilians argued all evening. Men who couldn't believe Napoleon would abandon Egypt shouted down those who dared to suggest his departure was imminent. The rumor gained strength, though, as various *savants* shared their own eyewitness observations. Geoffroy Saint-Hilaire had no doubt of it: he had told Napoleon of a draft manuscript he wished to post to a colleague in Paris. "Give it to me," he reported Napoleon to have said, "and it will soon arrive." Conté told how the general had three times that very day asked him to make a portrait of his mistress, Pauline Fourès. And Monge had just given Conté his personal provision of wine—what else could explain that? Another scholar reported that Monge had donated all his books to the Institute's library. There was also the giddiness of the geometer himself, noticeable for three or four days running, very curious.

Monge and Berthollet's return from the general's dinner silenced this conversation—temporarily. Sheepishly, the pair explained their packed luggage with the lie that they were going with Napoleon to visit the verdant delta province of Menouf.

Two days later, all pretense ended. Around six in the evening Napoleon received word that the Anglo-Turkish fleet had passed on, and the coast was temporarily clear. At nine, Monge, Berthollet, and Denon received orders to be ready to go, and they started moving heaps of baggage from their rooms to the gates of the Institute. At ten, a coach with a military escort arrived and the three men hurried out without a word, to the consternation of their watching colleagues, who were still afraid to voice their suspicions.

Finally an engineer called out, mockingly, "So, Citizen Monge, are we holding a meeting at the ruins of Thebes?" Monge, embarrassed, replied, "Yes, we'll meet at Dendara, err, under—over, Dendara."

"Are you passing through Damietta?" called out François-Auguste Parseval Grandmaison, the Commission's official poet. "I don't know," Monge replied. "I believe we are going to Lower Egypt." Under his breath he muttered something about "how quickly the general goes in his expeditions."

As the three men jumped into the waiting coach, the Institute secretary Fourier called out, "The Commission is alarmed at your sudden departure. Don't you want to reassure us, to cover our responsibility?" Finally, Monge stuck his head out the carriage window and confessed: "My friends, if we leave for France, we didn't know anything about it until midday today."

With that, the coach sped off toward the general's Mameluke palace, where Napoleon was saying goodbye to his Bellilotte and others in the garden. Some of the *savants* were also present, including Geoffroy Saint-Hilaire, who had been tipped off in advance by Monge. The naturalist later reported that Napoleon spent his last hour in Cairo engaging Monge and Berthollet in a discussion about physics and particle attraction, regaling them with his usual blather about the great scientist he himself might have been. "I find myself conquering Egypt as Alexander did; it would have been more to my taste to follow in the steps of Newton. This thought preoccupied me at the age of fifteen years," the naturalist quoted Napoleon. The general then presented a monologue on the magnificence of Newton's planetary discoveries, and expressed a wish to personally discover universal laws governing the smallest particles of matter. The hero-worshipping Geoffroy Saint-Hilaire recorded every word for posterity, and did not complain until much later about the betrayal of being left behind.

A little before midnight, the general and his party slipped out of Cairo by lantern light. They spent three days on the Nile, and finally saw their last of Egypt as they sailed from the coast in a small, unlit boat, the very opposite of *L'Orient*, on August 22. The poet Parseval joined them at the last minute, frantically rowing out to sea as the ship was about to raise anchor and crying to be taken aboard. Napoleon was inclined to leave the pleading scribe behind, but Monge intervened.

The party sailed for seven weeks, in constant fear of the English. Napoleon gave Monge the order to blow up the French ship if the British caught up to them, instructions the dutiful Monge took a little too seriously. When mysterious sails appeared on the horizon one afternoon, all hands took up positions to repel the attack. The ship turned out to be French, but Monge didn't reappear after the stand-down order. His shipmates finally found him in the powder magazine, a lighted lamp in his hand, preparing to blow the ship and all aboard into oblivion. Napoleon's surprise return cheered the populace in

France as much as it distressed the men he had abandoned in Egypt. Denon described the scene at Toulon: "Scarcely was the flag of the commander-in-chief on the mainmast before the shore was covered with people who uttered the name of Bonaparte; the enthusiasm was at its height, and it produced disorder; the contagion [plague] was forgotten. Sublime effusion!"

Back in Paris, the government was unstable; the Directory roiled with intrigue and backstabbing. In addition to the military disasters in Europe, there was financial trouble, not to mention propaganda pressure from the British, who had published a book of intercepted French letters from Egypt, with notes explaining that the French government had wantonly and deliberately sacrificed tens of thousands of suffering French soldiers. The left-leaning Parliament tended to agree, and legislators had publicly charged that the Directory had "deported into the deserts of Arabia the elite of our troops, General Bonaparte, and the cream of our *savants*, scholars, and artists."

Several members of the Directory were already planning their own coups d'état when Napoleon arrived back from Egypt. Napoleon signed on to participate in one of these plots, with the promise that he would share ultimate power. But when the leftist Council of Five Hundred refused to accept his terms—essentially a de facto suspension of the Constitution and a leadership consisting of three men, Napoleon amongst them—the general dispatched his soldiers to the legislative chambers, bayonets fixed. The lawmakers jumped out the windows, and a few hours later, on November 9, 1799, Napoleon graciously accepted appointment from the Council of Ancients as one of three consuls in a provisional government. As the most powerful of the three, Napoleon assumed the role of First Consul.

The French population reacted apathetically to the bloodless coup of 18 *brumaire*, as it became known, proving the revolutionary era was over. Napoleon established a new order in France that would last for the next decade and be marked by years of European war and cycles of victory and defeat for the French. As consul, one of Napoleon's first acts would be to come to an agreement with the Catholic Church, in a treaty called the Concordat, which gave religion a place again in post-revolutionary France.

Although word of the Abukir disaster had reached France, the French people were none the wiser about the dire state of affairs in Egypt. The gruesome details—plague, the disastrous siege at Acre, the army near mutiny—were not widely known. The Oriental adventure only added luster to the reputation of the rising young leader. Napoleon was a master of publicity and he spun the story of the Egyptian expedition decidedly to his advantage. Arriving in Paris, he immediately saw to it that medals were struck to commemorate the campaign, depicting him as Mercury flying home after another glorious triumph for France. The nation was also in such a state of need and disarray—with all of Napoleon's European military gains coming unglued and the government collapsing—that there was no time to analyze and uncover what had really happened to the 50,000-man French force thousands of miles across the sea. What mattered was that the great warrior was back in France, in time to defend the nation in its gravest hour.

Denon left Egypt with a mixture of joy and sorrow. "I was favored with the realization of a dream," he wrote of Napoleon's invitation to accompany him back to France. "But a sentiment for which I could not account made me miss Cairo; I had scarcely lived there, and yet I had quitted it with sorrow." He credited the mild temperature and dry desert air for the city's strange seductiveness, adding that it caused some Europeans, "arrived for a few months at Cairo, to have grown old in that city, without ever persuading themselves to leave it." Denon returned to his Paris apartment with his collection of Egyptian prizes, including a papyrus from a Thebes mummy, an entire mummified foot, and his sketches, and set to work preparing his drawings and notes for what would be the first nineteenth-century best-seller on Egypt. Ever the attentive ladies' man, Denon gave his patron, Joséphine, the whimsical present of a small African monkey specially trained to seal letters.

Monge and Berthollet arrived in Paris looking and smelling like beggars—they hadn't had a change of clothes since Egypt. Monge's wife's porter nearly didn't open the door for the bedraggled master. The duo quickly cleaned up and resumed their places at the Institute of France. Napoleon himself appeared at two meetings of the scientific body in late October. The scientists gave him a standing ovation. A week later he returned to talk about the Rosetta Stone and the great work French engineers were doing in Egypt, searching for a way to dig the canal.

Back in Cairo, soldiers and scholars alike reacted with dread, disgust, and despair to the news that their leader had fled. The army was approaching shambles. A month before he left Egypt, Napoleon had tacitly acknowledged this when he offered to pay Turkish prisoners held in the Citadel if they agreed to join the French army. The officer Napoleon left in charge of Egypt, Jean-Baptiste Kléber, had lost all faith in the general and the Egyptian campaign months before, and he made no secret of it. In one diary entry, he wrote of Napoleon: "He is incapable of organizing or administering anything and yet, since he wants to do everything, he organizes and administers. Hence our want of everything, and poverty in the midst of plenty."

Those scholars who most admired Napoleon were also the most disheartened when he left. For others, like Desgenettes, their already low opinion of the leader was simply confirmed by his cowardly flight.

One scholar who had lost his faith in Napoleon was the geologist Dolomieu, and he was already out of Egypt and meeting a terrible fate. Dolomieu had first offended the general by producing a report about the architectural remains at Alexandria in which he used the following epigraph: *Tempus edax rerum*. Needless to say, "Time erases all things" was not among Napoleon's favored sayings; but Dolomieu, a respected geologist known throughout Europe, wasn't capable of producing politically correct science. Napoleon had asked Dolomieu to survey the Egyptian soil for agricultural purposes, and the geologist's report deflated the general's hopes of creating a breadbasket on the Nile. Dolomieu determined, after doing fieldwork, that the delta couldn't produce much more food than it already was producing. Scurrying out of that meeting as the general glowered, Dolomieu had muttered, "I am saving myself from the lion's claws."

In January 1799, when Dolomieu requested permission to return to France on grounds of ill health, Napoleon gladly dispatched him. A bad storm and leaky boat forced the small group of travelers, which included the Herculean general Dumas, to dock in Sicily where the royalist Neapolitans promptly threw the French in jail. Among the captors were some exiled Knights of Malta, who were more than happy to get their hands on a man they viewed as a traitor to their Order, for negotiating the fall of Malta for Napoleon. The Brothers accused Dolomieu of treason and locked him in a damp, barely lit dungeon measuring ten by twelve feet, where he languished for nearly two years. During his imprisonment, he scratched out a groundbreaking geological treatise—using lumps of burned coal and writing in the margins of a Bible—called *The Philosophy of Mineralogy*. He was released only after strenuous efforts by the intellectual community in Paris, but he returned to France in such ill health that he died in November 1801, just as the last of his Egyptian colleagues were finally returning to France. He died a bachelor, in accord with his vows as a Knight of Malta, although, his biographer noted, "he was known to fancy women."

Some of the young scholars sobbed at the news of Napoleon's betrayal. "We reacted emotionally to this escape, but calmed down," wrote Devilliers. The general's departure crushed their spirits, but worse, they had lost three of their most important members.

With the British blockade intercepting the correspondence from home, the scholars felt ever more forgotten by colleagues and families. They wrote letters, but rarely received replies. A few weeks after Napoleon left, Devilliers wrote to his brother complaining that his family had not been writing. "I am bored to death, losing the best years of my life and health, and spending these years in sadness." He added that the rest of the young people all felt similarly. Their only hope was that with Monge and Berthollet back in Paris to plead for them, they too might be sent for sooner rather than later. Even letters that did manage to reach France were often unreadable when they were delivered. Devilliers sent another letter in September 1799 to his father that arrived completely illegible, thanks to the sanitary precautions taken at Marseille. In fear of the plague, French officials soaked all Egyptian correspondence in vinegar. The Devilliers family took the letter to Berthollet, who applied his skill as a chemist to help restore the writing. As it grew legible, they realized he was begging his father to make a personal pilgrimage to Monge and Berthollet to plead for the scholars stranded in Egypt. "I dream of my family, and I am sad to wake up from this imaginary consolation," he finished.

When Napoleon abandoned them, the scholars had been in Egypt for a full year. While they longed to go home, they had become relatively accustomed to the climate, culture, and deprivations. The Egyptians, however, had grown no happier with the French presence—on the contrary, they were simmering with anti-French sentiment. In March 1799, the army's chief administrator had noted, "The people are inflammable. They suffer the rule of a Christian nation so impatiently that it would only take a spark to touch off a general revolution."

The scholars were careful not to provide that spark, but 150 unarmed civilians could do little to influence public opinion when tens of thousands of French soldiers were committing myriad offenses daily. Devilliers described what was apparently a typical day of French atrocities during his mapping assignment in the desert. One morning, he wrote, the engineers and their French military accompaniment encountered an Arab family, consisting of a father, mother, and teenage son, and their camel. The French interrogated them as spies, while "the twenty-five soldiers made the poor woman submit to the most ignoble treatments," Devilliers wrote. Eventually, the soldiers shot the young man for failing to provide information. The image of the boy falling was permanently seared into the young engineer's mind, he wrote. The father then ran behind some bushes, but the soldiers captured and killed him, too. Later the same day, the French group met a herd of lambs shepherded by a girl of nine or ten. The soldiers killed and took as many of the lambs as they could carry, then put the girl on a camel and brought her to Suez, where, the young engineer wrote, "the general's staff sold her to a captain from Yambo. I returned to Suez, my soul upset with what I had seen." Jabarti also described routine rapes.

The *savants*, of course, interacted more peacefully with the Egyptians than did the military. They paid for their food, lodging, and donkeys or camels (rather than simply seizing them). They learned Arabic. Some, like Conté and Coquebert, became quite proficient; Conté because he needed to speak with the Egyptian craftsmen, and the young botanist-librarian because he met so many educated Egyptians in his library. Sometimes, they tried to intervene to prevent perceived injustice. Dubois-Aymé and Jollois, for instance, wrote of trying to help peasants being beaten by tax collectors in the delta town of Menouf. They were lodged in a large house with a Coptic *intendant* (a steward or paymaster) who lived in the lower part. From their windows they could see him routinely beating peasants who did not voluntarily pay the tax. The young men insisted they would intercede for the poor, while the Copt argued that the *fellah* would not pay unless constrained to do so on threat of pain. "We often saw men who had been uselessly beaten many times finally pull from their mouths, or from the folds of their turbans, the money asked for and give it to the collector," they wrote.

The young scholars felt they were bringing a certain honor to the French image in Egypt by their acts. They believed the Egyptians appreciated their compassion and acknowledged it as a French trait. "It is unimportant to be named, as long as it is said 'a Frenchman did that for me, a Frenchman helped me in this fashion,' " they wrote.

The scholars couldn't always act on their instinct to intervene in the harsher aspects of local culture. A few months before leaving Egypt, while traveling with soldiers in the desert outside Alexandria, Denon met a recently blinded woman, blood still dripping down her eyeless face, clutching an infant and begging for food. When the French unit stopped to offer water, a man claiming to be her husband trotted up and demanded that they stop giving aid. "She has lost her honor," the husband shouted, according to Denon. "She has wounded mine; this child is my shame, it is the son of guilt!" The horrified French artist watched helplessly as the man drew a dagger, stabbed the woman to death, and then hurled the infant to the ground, killing it as well. Denon asked some Egyptian guides whether such brutality wasn't against the law. The man was legally within his rights, the guides explained, although local custom frowned on the murder. "The man had *done wrong* to stab the woman," Denon was informed, "because at the end of forty days, she would have been received into a house and fed by charity."

Witnessing spousal murder and infanticide never diminished Denon's enthusiasm for Egypt. He scorned Egypt's detractors among the French army. "They wanted gold," he wrote, "and not finding gold, saw nothing around them but burning sands, fleas and gnats, dogs that disturbed their sleep, intractable husbands, and veiled women who showed nothing but the eternal neck!"

Most of the French who witnessed such incidents passed quick judgment on Egyptians in general, describing, for example, the Arab character as "inscrutable" and filled with *"sangfroid."* The French observed the intense, otherworldly spirituality of deeply religious Muslims, but they didn't understand its source. "Nothing disturbs them," wrote one French officer of the Egyptians. "Death to them is like a voyage to America for an Englishman."

The French were appalled at the overt misogyny they witnessed, but many of them also indulged in the local practice of female slave-holding. Bernoyer chronicled the countless hours and much coin he spent trying to find the perfect odalisque to keep him company in Egypt. Some French were impervious to the charms of the Egyptian female. After months in Egypt, one disappointed scientist, despairing of finding Savary's fabled exotic nudes on the riverbanks, wrote, "We saw buffaloes in the water, but the Egyptian women were dirty as snails and black like moles."

Other Frenchmen actually found true love with Egyptian women. One French captain, Joseph-Marie Moiret, became entranced by a veiled beauty he met in the Cairo marketplace. By bribery and persistence he managed to find out her name and soon was "wafting kisses" to her surreptitiously, which she accepted. He learned her name, Zulima, and that she was a Caucasian harem woman who hailed from Georgia. Moiret romanced her throughout his stay in Egypt, but was forced by his military masters to leave her behind when he returned to France, promising to try to send for her but knowing full well that her fate, as the lover of an infidel and French occupier, would be terrible. He never saw her again.

French occupiers could equal Mameluke tyrants in brutality, but the French were atheists and voluptuaries, lovers of wine and women. Mameluke culture in late-eighteenth-century Cairo was strongly religious, Spartan without the asceticism (ruling the city for so long, they had become less self-denying than their forebears in eighth-century Baghdad), and deeply misogynistic. European visitors like Volney were always appalled at the Mamelukes' preference for sex with boys. "They are above all addicted to that abominable wickedness which was at all times the vice of the Greeks and of the Tartars, and is the first lesson they receive from their masters," he wrote. "It is difficult to account for this taste, when we consider that they all have women, unless we suppose they seek in one sex that poignancy of refusal which they do not permit the other. It is, however, very certain that there is not a single Mameluke but is polluted by this depravity, and the contagion has spread among the inhabitants of Cairo, and even the Christians of Syria who reside in that city."

Until forbidden by religious decree to do so, female Mamelukes in the Middle Ages dressed like men in order to attract men. Mamelukes did have wives and harems, but the position of women in their society was an aberration within the broader Islamic world. Female virginity, so highly prized in Islamic society, didn't matter much to them. Children did not inherit nobility from their biological fathers, so female fertility was not as important as in other communities. Mameluke women notoriously avoided childbirth to keep themselves fit and sexually desirable, and so routinely practiced abortion. If a Mameluke had a private taste for female virgins, he could buy them by the dozen. But he could only pass on his nobility to a boy who had been purchased from someone and raised in the traditional warrior style.

Partly because they were not as sexualized as other women, and partly because their men deemed matters of business and real estate to be effete occupations, some female Mamelukes (purchased from the Caucasus, like their men, never Arab) wielded considerable financial power. The wives of the wealthiest Mamelukes handled business not just inside the house, but beyond. Together, Mameluke men and women tax-farmed, accumulated lands and wealth, and occasionally returned their loot to the people via religious charity to the poor.

Female Mameluke history in Cairo is rife with tales of plots, poisonings, and court intrigue. One of the last and most famous Mameluke women in Cairo was Nafisa Al-Bayda (sometimes called Nafisa Khatun), the wife of the chief Mameluke leader Murad Bey when the French arrived in 1798. Nafisa was an imported slave herself, of unknown origin, from Anatolia or the Caucasus. She was reputedly both beautiful and sophisticated. Although no pictures of her exist, her nickname, "the White Jewel," gives some idea of how she appeared.

She could read and write in Arabic and Turkish, spoke French and English, and became a negotiator between the French, the British, and the Mamelukes.

A Mameluke legal document from 1791 refers to her in the most floridly deferential terms, owing to her powerful position: "Exalted among the veiled, glory of honorable ladies, crown of the illumined ones who carry the sublime veil behind the forbidding curtain of seclusion, the splendid hidden jewel, the flowering well-guarded gemstone: Lady Nafisa." A public fountain in Cairo that she commissioned in 1796 is decorated with a motif of breasts and a heart, in recognition that the benefactor was female.

Nafisa's path to power was typical for a Mameluke woman. Her first Mameluke master was a bey who purchased Nafisa from her family or from kidnappers. He was so taken by her beauty and wit that he eventually married her, putting her in control of vast real estate holdings. When Murad Bey killed her first husband and then married her, she brought to the union an entourage of more than fifty women and several eunuchs, not to mention acres of property, palaces, and commercial ventures. Nafisa freed and married off many of her own slave women into grand households. Through these freed female slaves, she had her own power network within the Mameluke hierarchy.

When the French arrived, Nafisa was a formidable member of Cairo's elite—even more so because her husband had left the city to fight the French in the desert. During the occupation, Nafisa remained so rich that French demands for large amounts of gold from her treasury barely dented her fortune.

Mameluke women like Nafisa had few, if any, children, preferring the role of power-wielding businesswoman or seductress to motherhood. When they did have babies, their progeny, although born in privilege, were not guaranteed military nobility, because they were not Mameluke slaves. Under their own rules, Mameluke children could never join the fighting elite to which their fathers belonged. That honor was reserved strictly for imported slave boys. Mameluke children played little or no role in Egyptian society other than as members of a parasitic elite and never assumed significant civic powers.

By the time the French arrived, the Mameluke system was dissipated, byzantine, and corrupt. After living in Cairo among the Mameluke clans for some time, Geoffroy Saint-Hilaire believed some beys had tried to artificially transform their children back into slaves in order to grant them Mameluke social primacy. "They sent their children far away, and had them sold by foreigners, and then bought them with the intention of giving them more consideration and to raise them to dignities," the naturalist reported in one letter home. Other French scientists who surveyed the modern state of Egypt reported that Mameluke children were weak and unable to thrive in Egypt. They compared them to foreign plants unable to take to the local soil. "Almost all their children died very young, and their race, it is said, rarely arrive at the second generation," they wrote. Whether their children died young, were sent away to be repurchased, or were simply never born at all, the Mamelukes of Cairo were a ruling class that perpetuated itself without relying on offspring.

The Mamelukes developed stringent rules to maintain the purity of their sect and keep power and wealth in their own hands. To preserve their separateness from the native population, the Mamelukes took Turkish first names and spoke a Turkish dialect. They used interpreters to communicate with the Egyptian people, and refused to speak Arabic amongst themselves, even though it was the official language of the region and of their religion, Islam. Only Mamelukes could own other Mamelukes, and Mamelukes were qualitatively different from black African slaves. Rich Arabs and Christians in the community bought Negro slaves, and Mamelukes might have African slaves in their households, but no Arab would ever own a Mameluke in eighteenth century Cairo—nor had for hundreds of years.

Only Mamelukes were permitted to ride horses in Cairo. Arabs and foreigners were limited to donkeys. As in other Islamic countries, Jews and Christians in Mameluke Egypt had to wear distinguishing colors and turbans of prescribed length.

The Egyptian Mamelukes could be cruel and greedy rulers, but they were also deeply attached to Cairo. They left it only to do battle, and stayed in their palaces even during the frequent plague epidemics, which hit them much harder than the native inhabitants. During the Black Death in the fourteenth century, for example, they refused to flee Cairo, and entire garrisons perished of disease.

Their palaces were indeed such pleasure domes as English poet Samuel Taylor Coleridge imagined in his "Kubla Khan" opium fantasy. Walled compounds, lavishly tiled with lapis and gold, they were oases of luxury in the tumultuous city. Fountains sparkled in courtyards, surrounded by gardens of jasmine and orange. They designed their mansions with confusing layouts, secret passageways, and hidden rooms in which to secrete treasure, women, and eunuchs. Some compounds were self-contained villages with storage areas, granaries, shops, stables, several kitchens, and numerous steam baths.

Mamelukes lived together in these "households," where dozens of purchased white boys studied religion and military arts. The boys called each other "brother" and called their master "father." When they were old enough, their masters officially set them free, and the grown men became part of the Mameluke power network, as attached to their masters and their caste as grown men are attached to fathers and family. Once freed, they grew beards to symbolize their freedom.

By the summer of 1799, the *savants* were going native. Many let their beards grow in the Muslim fashion. In a letter to Cuvier, Geoffroy Saint-Hilaire explained that beardless men were assumed to be slaves, so he and the other French *savants* stopped shaving. Some also took up the water pipe. Published drawings of them at work, nevertheless, show them always clean-shaven and in frock coats, perhaps in a nod to European sensibilities.

More shocking to Europeans, some indulged in the local custom of owning slaves. Geoffroy Saint-Hilaire himself had two—a boy and an old woman. In letters home, he claimed to view them as family members. "I am living here very peacefully, busying myself by turns with natural history, my horses, and my little black family, to whom I have momentarily transported the tenderness that is useless to my European family," he wrote to his father in June 1799. "I bought an eleven-year-old child for 250 francs whom I trained to care for my collections and to stuff animals. Since then, I was given an old Negress who is very skillful at housework. I have a donkey for my faithful Tendelti, the Negro, who is eleven years old."

Knowing such news would raise eyebrows back home, where slavery was abhorred, Geoffroy Saint-Hilaire elaborated. "Slavery here is different than in America. It is a veritable adoption. My two slaves never call me anything but their father, and I am so satisfied by their services that I feel the same friendship for them." In the same letter, he also ruminated on the oddly inverted position of the slave in Egypt, noting that in Mameluke society, slaves were more honored than the free.

Geoffroy Saint-Hilaire was hardly the only Frenchman who indulged in slave holding. The chief military engineer Girard bought himself "a Caucasian woman" for 3,600 livres. The military chronicler Bernoyer devoted page after page in his memoir about his and other Frenchmen's efforts in Cairo to locate and purchase the ideal female slave. In April 1799, Devilliers watched French soldiers pillage a Darfur caravan on the pretext that it carried weapons to be used against the French. The soldiers who participated in the raid rewarded themselves with the caravan's booty, including precious goods, camels, and male and female black slaves. They then sold the slaves on the open market, pocketing the price.

General Kléber, whom Napoleon left in command when he sailed back to France, was tall, blond, muscular, of French-German stock, and already deeply disillusioned with the Egyptian campaign. Napoleon, afraid to inform Kléber of his new job in person, dispatched the unwitting general on a pointless mission to Rosetta, then sent Kléber a commission letter with instructions as he was leaving Cairo.

Kléber was an odd choice for a replacement among the twenty-seven generals Napoleon had brought with him to Egypt. (A general's job was exceptionally perilous—two would be assassinated, three mortally wounded in battle, nine wounded but survive, and two die of illness before it ended.) A literary man, he was trained as an architect, a hero of the Italian campaign, he had lost all faith in Napoleon at Acre, and was livid at Napoleon's escape to Europe. Handed the letter, he couldn't contain his rage. "That bugger has left us with his breeches full of shit!" he shouted after scanning it. "We'll go back to Europe and rub them in his face!"

The trouble was, how to get back to Europe to realize that satisfaction? In the letter, Napoleon empowered Kléber to surrender only under certain conditions. "If by next May, you received neither help nor news from France, and if, in the coming year, despite all precautions, the plague should kill more than 1,500 men, you are authorized to make peace with the Ottoman Porte, even if the evacuation of Egypt should be the principal condition."

Napoleon filled his letter with platitudes regarding fortifications, hospitals, keeping on good terms with the Muslims, and remembering Alexander's legacy. He did admit that "The plague is one of the army's most redoubtable foes." He failed to mention that the army was now penniless, mutinous, and sickly, and that the Turks were sending 80,000 men to expel them from Egypt.

Kléber, along with a sizable number of the other French military leaders, wanted to conclude the Egyptian adventure with the quickest and least disgraceful surrender. Other "colonialist" generals still hoped

Egypt could be turned into France's jewel of the Orient—a tourist attraction at the very least, a commercial center to rival British India at best. Among the latter camp was General Menou (full name: Jacques-François de Bussay de Menou). Menou became more committed to colonialism after he took a Muslim wife, converted to Islam, and renamed himself "Abdullah." Menou would have been a more logical choice to take command of Egypt after Napoleon. Napoleon probably chose Kléber over the colonialist Menou because Kléber was much more well-liked by the soldiers and Napoleon was trying to stem an open split in the military in his wake. In any case, Kléber was forced to make the best of a bad situation.

Kléber was indeed beloved by the soldiers and the scholars. A soldier's soldier, his first concern was for the troops, and he laid a brutal tax on the Egyptians to keep the French fed and clothed. The civilians respected his intellectual side and elected him a member of their Institute. Kléber was "fair, brave, and modest," Devilliers wrote. "At first he did not want to be part of the Institute, asking in what area he would be placed. But he finally agreed, saying, 'Put me in the arts, it is in that that I get along the best.' "

Reluctantly taking charge of the Egyptian campaign, Kléber tried to negotiate a face-saving surrender with the Turks and English, while still making sure the sickly and undersupplied French army survived and left with honor. To accomplish the latter, he was much harder on the Egyptians than Napoleon had been. He reentered Cairo after Napoleon's departure with great pomp, preceded by a double column of 500 janissaries, who struck the ground with their poles and shouted, "Muslims, bow down before him!" Turkish janissaries in Cairo served as part of the French occupation's police force. The Turks based in Cairo had fluid loyalties. The crushing taxes he levied on the Egyptians inspired a hatred that eventually led to his demise. In spring 1800, Cairo revolted again, this time for thirty-seven days. Kléber crushed the uprising even more brutally than had Napoleon the previous year. French artillery smashed all the palaces around the Ezbekiyah Square. Jabarti described the aftermath in a single mournful sentence. "All turned to sparkling fire and ruins, as though there had never been enchanting villas here, or gatherings of friends, or promenades."

The scientists had dared hope that Napoleon's departure meant their own escape was imminent. They were wrong. As disgusted as Kléber was with his predecessor and the situation in Egypt, he had his own reputation to think about. The scientists weren't going anywhere.

Napoleon had empowered Kléber to send the scientists home when their work was finished, adding that if Kléber thought any of the scientists might be useful to him, he could keep them in Egypt. Kléber oversaw the scholars' fieldwork among the ancient sites in Upper Egypt, and ordered them to begin studying the modern state of Egypt. Kléber also first suggested they compile a book. The book ultimately became a lasting monument to their work, but at the time, the scholars could think only of how Kléber's orders delayed their return to France. Around this time, Geoffroy Saint-Hilaire complained bitterly about being used as a pawn by legacy-minded generals. "The poor *savants* of Cairo were thus brought to Egypt in order for one to be able to read in the history of Bonaparte another line of praise, and they are kept in order that in that [the history] of Kléber there is no reproach," he wrote. "Thus, the small have always been the toys of the big."

The army had a new leader, and so did the scientists. With Monge and Berthollet gone, Joseph Fourier took charge of the Institute. Fourier, though eminently capable, was not terribly popular among his peers, yet he was proving himself to be an able administrator and diplomat for the French army. He was also a scientist of genius. As with Malus and Monge, the desert climate inspired Fourier to think in new ways. In Egypt he first became interested in the conduction of heat. His later theories on that subject are so important, they have been described as forming "a trunk nerve of mathematical physics."

Comte Claude Louis Berthollet, 1748–1822. Chemist and co-leader with Monge of the *corps des savants,* he helped create the table of elements. After Egypt, he founded the Society of Arcueil, which counted among its members many of the prominent European scientists of the early nineteenth century.

Nicolas Conté, 1755–1805. Chemist, painter, and brilliant inventor, of whom it was said, "He holds all the arts in his hands and all the sciences in his mind." Among his many innovations were new gas combinations for hot-air balloons (in the search for which he blew up his laboratory and lost an eye). His most enduring invention is a method of mixing graphite and clay that allowed the mass production of pencils without great quantities of lead, a process still in use today.

Dominique Vivant Denon, 1747–1825. Artist, diplomat, author, pornographer, Denon was appointed first director of the Louvre Museum by Napoleon after the Egyptian campaign.

René-Nicolas Desgenettes, 1762–1835. Doctor to plague victims in Egypt, aristocrat, wit, he served as chief physician for the French army throughout the Napoleonic years.

Joseph Fourier, 1768–1830. A genius and early physicist, whose theorem on waves is still taught in basic calculus, Fourier initiated groundbreaking studies on the properties of heat after his years in the desert.

Étienne Geoffroy Saint-Hilaire, 1772–1844. Zoologist, professor, zookeeper, he proposed a controversial theory of unity of life that, in its provable aspects, played a role in the development of the theory of evolution, and in its more mystical aspects, made him a hero to Romantic literati, including Honoré de Balzac and George Sand.

Jean-Baptiste Prosper Jollois, 1776–1842. One of the engineering students whose rigorous methods of measuring and drawing the ancient ruins in Egypt set the standard for archaeologists who followed.

Gaspard Monge, Comte de Peluse, 1746–1818. Co-leader with Berthollet of the *savants,* Monge was a peddler's son and mathematician. He invented descriptive geometry, a new way of mathematically depicting space, which enabled engineers to design the machines that powered the Industrial Revolution.

Marie-Jules César Lelorgne Savigny, 1777–1851. Botanist turned zoologist, his meticulous work on the similarities between insects played a small but important role in the development of the theory of evolution, but left him blind.

Expecting to find vestiges of ancient glory, the French were disillusioned at their first sight of Alexandria. Soldiers and scholars alike sheltered themselves from the burning sun under makeshift palm huts.

The Sphinx was up to its chin in sand when the scholars inspected it. They knew there was more to it underground, but they had neither time nor machinery to excavate it.

The French looked along the Nile for Cassas's beautiful bathing women, but found instead only "black women, dirty as snails," complained one soldier.

The Mamelukes ritualistically dressed themselves in layers of silk, and carried all their wealth on their bodies.

The last of the Mameluke beys, Murad was an aesthete, who preferred music and literature at his palace near Giza to the intrigue of Cairo court life. He led the French on a wild chase up the Nile, giving artist Denon his first glimpse of Egypt's greatest ruins.

In Cairo, the French made their headquarters in the medieval fortress called the Citadel.

Napoleon took over one of the finest houses in Cairo, on the elegant Ezbekiyah Square, which was turned into a lake during the annual Nile flood.

The young students were shocked at the brazen dance moves and suggestive lyrics of the dancing girls called *a'lmes*, in Cairo's streets.

The naked Muslim holy men the French called *santons* danced in the streets until they dropped, and took unbelievable liberties with women.

Monge and Berthollet took possession of one of the finest mansions in Cairo, recently built and then abandoned by a wealthy Mameluke and his retinue.

The bey's garden was tranformed into a retreat of enlightenment. Geoffroy's zoo, a printing press, and Conte's workshops were installed here, as well as a few scholars.

meetings and read papers to each other in an abandoned harem room.
comes Napoleon to a meeting, and Geoffroy and Berthollet follow the

is enraged when
pt and put him
t bugger has left
es full of shit," he
t back to Europe
s face."

...odified his invention, the flying ambulance, to fit camels.

A Muslim fanatic stabbed the French general Kléber to death in a Cairo garden. The French responded with what they thought of as Muslim punishment—chopping off the killer's hand and then impaling him.

...e fascinated by the pristine condition of the ancient temple of
...essed with the zodiac, believing it held clues to the ancient civiliza-
...of astronomy.

Etched by J. Gillray, from ye Original Likeness by Denon.

nd-
ion
rs'
ou-
cat-
as
es.

gry
of
y
re.

Geoffroy dissected and sketched the Nile crocodile, but he
...sted his son and biographer, actually tried to tame one, British
...ithstanding.

...aptured a number of these Egyptian mongooses and brought
... of them back to France.

Geoffroy was well on his way to discovering his universal life theory when he opened the rays.

Bats—"flying mice" in French—tormented the scholars as they descended into the gloomy temples and tombs.

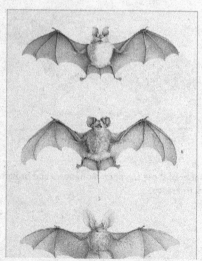

Savigny had his greatest harvest of creatures on the shores of the Red Sea.

No insect was too small for Savigny's meticulous eye. He drew separately the individual mouth, body, and leg parts, as well as the entire creature.

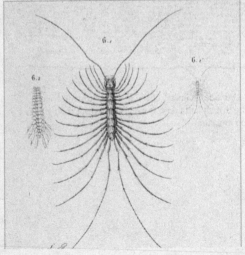

re able to include an image of the Rosetta Stone in their
hough the British had taken the real thing from their hands
a London museum.

Fourier's best-known work was already behind him when he got to Egypt. His theorem on the mathematical underpinning of wave energy is still taught to calculus students as Fourier's theorem. He had also discovered the Fourier series, another fundamental building block in calculus. His work had broad and lasting impact. "Fourier's theorem is not only one of the most beautiful results of modern analysis, but it may be said to furnish an indispensable instrument in the treatment of nearly every recondite question in modern physics," wrote the nineteenth-century British scientist William Thompson (Lord Kelvin).

Fourier was the orphaned son of a poor family and would not have risen far in prerevolutionary France. His father was a tailor in Auxerre, a town southeast of Paris. Both parents died before he was eight. Put under church care, the local bishop noticed Fourier's studious bent and placed him in a Benedictine military college, where by the age of twelve he was writing sermons that dignitaries in Paris read as their own. By his early teens, he proved also to be a math whiz. Legend has him collecting candle ends in the college kitchen to study at night in a screened inglenook.

Fourier was working toward the priesthood in 1789 when the Revolution ejected the clergy, and he never regretted losing that career option. By age twenty-one, Fourier was in Paris, writing and presenting papers on mathematical equations that advanced the work of older, leading scholars.

Enthusiastic about the Revolution, Fourier was repelled by the Terror. He hid out in Auxerre during the worst of it. Even so, he was arrested as a supporter of the radical Robespierre and avoided the guillotine only after his intellectual colleagues pleaded for him. In 1794, in a calmer political climate, Fourier was tapped as chairman of the mathematics section at the new École Polytechnique and given the forbidding title "Supervisor of Fortification Lessons." He was twenty-four years old.

Fourier excelled as a professor. The revolutionary government wanted a new style of educator, more invigorating than the old, and the actual creators of new mathematics were hired to teach it. Professors were ordered not to speak from notes, to the benefit of Fourier, who had a talent for extemporaneous lecturing. His lectures were not dry mathematical drones, but lively and sprinkled with frequent historical allusions and entertaining real-life applications.

Childhood deprivation had molded the orphaned Fourier into a cool and somewhat callous adult. Keen with numbers and spatial relations, he was also dyspeptic and insomniac, sarcastic and lacking in collegiality. Desgenettes thought the hyperlogical young man was "missing a bit of character."

Geoffroy Saint-Hilaire liked him even less. The two men's personalities were opposite: Fourier was dry, cool, intrinsically devoted to the explanatory power of numbers, while Geoffroy Saint-Hilaire was sentimental, easily excited, and more speculative in his approach to science. The two men's spiritual and professional differences grew starker in the relatively small scientific community in Cairo, and their personalities clashed during the months and years in Egypt. Typically, though, Geoffroy was the only one of the two who recorded his feelings.

In letters from Egypt, Geoffroy complained about Fourier's sarcasm and arrogance. "We were so close together, and his claims became so overweening, that we often came in conflict," he wrote to Cuvier. The men managed to forge a working relationship, but Geoffroy Saint-Hilaire never got used to Fourier's arrogance, or his assumption that the mathematicians and engineers were simply smarter than the naturalists and artists. "Fourier's aim was to have a name for the same superiority of understanding which it is customary to afford in Paris to [mathematicians Louis] Lagrange and [Pierre-Simon] Laplace," the naturalist wrote.

Fourier was complicated. His arrogance was balanced by better qualities, which made him an able negotiator and administrator with the Egyptians. The science biographer Arago, who met him years later,

lauded Fourier's "gentle manners, scrupulous caution in terms of preju-
dice, and mind of inflexible justice [which] gave him an influence over
the Muslim population." The engineering student Jollois, who lived
with Fourier in Egypt and was in a position to know him well, wrote
that the mathematician "knew how to conciliate many different opin-
ions by the agreeability of his mind, his urbanity, and his kindness."

Kléber had befriended Fourier during the Mediterranean crossing,
and he so trusted Fourier's abilities that he made him his official inter-
mediary with the Mamelukes and Egyptians. Fourier was thus the chief
French negotiator with "the White Jewel," Nafisa, when the combined
Turkish and British force loomed against the French. Through Nafisa,
Fourier passed word to Murad Bey that the French would give Murad
the governorship of Upper Egypt in exchange for his forming an alliance against the Turks. Murad accepted, and after 1800 the French and
the Mamelukes were allies. The alliance came a little late for Cairenes,
though. Murad Bey had intercepted a food convoy of 4,000 sheep
shortly before Fourier's offer reached him, and the city was starving.
The deprivation was the spark that provoked the April insurrection.

Fourier's diplomatic missions were sometimes harrowing. Sent to
negotiate a surrender with the Turks after one battle outside Cairo,
Fourier wound up not in the harem quarters—the customary quiet
place usually selected for such discussions—but in a house half-ruined
by cannonballs, in the middle of a quarter still rocked with fighting.
Just as Fourier was about to celebrate with coffee, per the local custom, the arrival of the Turkish Commissioner, gunfire exploded from
the house across the street. A bullet sailed through the coffee pot the
mathematician held in his hand.

Fourier might not have been beloved by his fellow scientists, but he
was one of the busiest. One of Napoleon's last acts before leaving Egypt
was to appoint Fourier to head two scientific expeditions to explore
Upper Egypt. In addition to that assignment and his negotiating tasks,
Kléber asked Fourier to commence a survey of modern Egypt. When

he wasn't carrying out his duties for the government or the Institute, he wrote his own esoteric papers on wind-activated watering machines and algebraic equations. He produced at least four mathematical papers at Cairo. He also participated in field expeditions to explore everything from Egyptian antiquities to the alkaline desert lakes.

Fourier didn't feel it at the time, but the Egyptian climate and deprivations of the siege worked a strange metabolic transformation in his body. When he finally returned to the northern latitudes, he felt perpetually cold. Without a name for his condition (modern doctors might diagnose myxedema, a thyroid disorder), Fourier decided that desert heat was simply the healthiest climate for a human body and that a dose of heat would cure all that ailed him, including his disordered heart.

For the last three decades of his life, he tried various means to recreate in his own offices and rooms in France the dry, baking warmth of a Cairo afternoon. He wrapped himself in wool throughout the year, and heated his tiny apartment with so many stoves that visitors could barely spend ten minutes with him before needing to step outside and cool off. Hotter than the Sahara and hell combined, his friends complained.

Fourier's greatest scientific work after Egypt had to do with theories about the physical nature of heat. While still in Egypt, he began making calculations that led to his groundbreaking study, *Théorie analytique de la chaleur* (Analytic Theory of Heat). It won the French Academy's Grand Prize in 1822, in spite of what some pure mathematicians saw as a lack of rigor. Fourier's study focused on a subject of great personal significance to him: the phenomenon of radiation, the action of the thermometer, the heating of rooms. Fourier never apologized or even answered his critics: he was, in fact, practicing a new kind of mathematics, applied mathematics, and he is remembered as one of the world's first mathematical physicists. An English scientist who later worked on the same heat problems, Lord Kelvin, called Fourier's theory "a great mathematical poem."

Once the shock of Napoleon's abandonment wore off, the scholars returned to their collecting, categorizing, drawing, mapping, and measuring. Without fresh supplies from Europe, and without Napoleon's influence on the army supply masters, they relied on salvage, severe economies, and Nicolas Conté. The inventor was now indispensable to the functioning of both the military and the scientific efforts in Egypt. Kléber routinely addressed him in letters as "my very dear and precious Conté," and once wrote, "You know how much I am attached to you." Conté's greatest sorrow was that because he was so indispensable in Cairo, he was never allowed to roam Egypt as freely as his colleagues.

Kléber waited just four months after Napoleon's abandonment to negotiate a surrender with the Turks mediated by the British in the person of Sir Smith. Besides dissension in the ranks, and disease, the French were under tremendous military pressure from the two enemies by this point. Eighty thousand Turks had marched down from Syria, some led by the Butcher himself and others by the Grand Vizier, leader of the Turks from Constantinople. They had already retaken Gaza and reached Al-Arish, where they massacred surrendering French troops.

The British, still reluctant to confront the French in a land war, were secretly fomenting discord among the French ranks, and were circulating documents among the French troops urging soldiers to demand an immediate return to France. French soldiers were mutinying in Alexandria, rioting to prevent the departure for France of several officials. "Either we'll all die together or we'll all get out together," they shouted, according to one witness.

Under the terms of surrender, signed in late January, called the Convention of Al-Arish, the French would evacuate the northern delta towns within ten days of the ratification and then leave Cairo (toward whose gates the Turks were fast approaching) within a month. The Turks would then ferry the French home, with their arms, at Turkish expense. Unfortunately, the British government decided not to honor any aggreement between the Turks and French, and in early March word reached Sir Smith that the French should be captured at sea and treated as prisoners of war. Smith tried to buy time in order to persuade his superiors in London to change their minds, but the Turks were bent on retaking Cairo immediately, as per the agreement. Kléber was left with no choice but to fight.

Called in from their various collecting expeditions, the scientists were ecstatic at the prospect of going home. They loaded themselves and all their collections onto dhows from Cairo, and onto a ship at Alexandria in spring 1800, only to learn after weeks floating in the harbor that the British had refused the French-Turkish terms. Kléber valiantly led the French against the Turks, and recaptured swaths of northern Egypt within a week by winning a famous battle against the Turks at Heliopolis. The battle was memorialized in coins and legend, but Kléber was a reluctant, if not remorseful, victor as he stepped back into the role of commander of Egypt.

In June 1800, a Muslim fanatic from Syria ambushed Kléber while he was walking in a Cairo garden and stabbed him to death. The assassin was caught and soon confessed. He was punished grotesquely, in what the French considered "Muslim style." They burned off his murdering hand to the elbow and then slowly impaled his whole body on a stake. Three members of the Cairo Divan were also executed—beheaded—for allegedly conspiring with the killer. French observers were horrified at the public torture. "Apparently, it was atrocious," Devilliers (who was not in Cairo) later wrote. "His agony lasted more than three hours. Even soldiers pitied him; one of them helped him to die by giving him something to drink in spite of the Chief of the Mamelouks." Ever the avid collectors, the French saved the assassin's corpse and brought his skeleton back to Paris, where it is conserved in a museum, missing three vertebrae and the bone of the severed right arm.

Conté undertook to arrange Kléber's funeral arrangements. The "indispensable man"'s last job for Kléber was to design an airtight lead coffin in which to send the general's remains back to France. Fourier gave the eulogy, a long-winded paean that, depending on which historical account is consulted, either reduced the mourning soldiers to a torrent of tears ("his voice was lost among the sobs," wrote Arago) or was full of "hollow bombast."

In his speech, Fourier called Kléber "a friend to soldiers" who "saved them bloodshed and lessened their sufferings." He implied that Kléber had died for the troops, paying with his life for the harsh taxes he had imposed on Egyptians in order to feed them.

"It is true that the troubles of the army were his daily concern and he thought of nothing but how to ease them," Fourier said. "How terribly he was distressed by the delay, then inevitable, in paying the army, despite the special taxes which were the object of the only harsh measure he ever gave. He devoted himself to regulating the finances of the army, and you well know how successful he was."

Without Monge and Berthollet in Egypt, and with Kléber dead, the scholars' relationship to the restive military grew ever more strained. Napoleon had been their number-one patron, and his departure had left them unprotected, subject to increasingly open ridicule and abuse. Kléber's protection was now gone, too. As the scientists' collections grew, word got around that they were collecting treasure. With food and money scarce, soldiers began eyeing the scholars and their trunks with vulpine interest. For the rest of their stay in Egypt, the scientists worried more about French soldiers raiding their collections, than about being robbed by Egyptian bandits or ambushed by military foes.

"Our situation has not gotten better since Bonaparte's departure," Geoffroy Saint-Hilaire wrote to Cuvier. "We were before welcomed everywhere with disfavor and ridicule. Bonaparte knew how to restrain his army with regard to us, and consoled us for the vexations we endured, telling us sometimes that however much the soldiers joked about the *savants*, they esteemed us. Today it only remains for us to envelope ourselves in our coats."

The zoologist knew the scholarly collections were provoking the soldiers' envy and suspicions. "We have gathered the materials of the most beautiful work that a nation could have undertaken. When we shelter so many precious riches from events, we are afraid to excite the jealousy of the soldiers." He ended his letter predicting that the scholars' final work would "excuse in the eyes of posterity the thoughtlessness with which our nation precipitated itself in the Orient." He begged Cuvier not to share his fears with his father.

Desperately homesick, suspected by the soldiers of hoarding treasure, they kept faith with the belief that their exile had a greater purpose. Their only hope of getting home soon, they now believed, was to complete their work. With makeshift tools in hand, they wrapped their burning heads in linen, hired boats, donkeys, and horses, and headed out on a series of field expeditions the likes of which had never before been attempted in Egypt.

CHAPTER 8

THE ARTIST ◿◿◿◿◿◿◿◿

> This abandoned sanctuary, insulated by barbarism and
> returned to the desert . . . was still a phantom so gigantic
> in our imagination, that the army, at the sight of its scat-
> tered ruins, halted of itself, and by one spontaneous impulse,
> grounded its arms, as if the possession of the remains of this
> capital had been the object of its glorious labors, and had
> completed the conquest of the Egyptian territory.
> —Denon, on first sighting Thebes in March 1799

Upper Egypt, Fall and Winter 1799–1800

When the French arrived in Egypt, centuries of dust
and debris shrouded the ancient Egyptian temples. The
ground level at Thebes and Karnak was several meters higher than
today. Sand had drifted to the tops of columns, so that only the
carved lotus and papyrus cornices, and the occasional colossal fist or
head, were visible. The Sphinx was buried in sand to its chin. Statues
lay exactly where ancient marauders and time had laid them low. In
unopened tombs the pharaonic paint—ochre, periwinkle, red—was
still bright. The epic excavation and plundering of the Nile that inau-
gurated modern Egyptology hadn't begun.

As time passed, the scholars grew more curious about these relics
than almost anything else in Egypt. Older than the oldest antiquities in
Europe, bigger than anything the Greeks and Romans had left behind,

their age, size, and design posed endless anthropological, architectural, and historical questions. What religion, what political order, what manner of men had erected these monuments before recorded time, and why?

Ruination deepened the mystery. The defiled faces of the Egyptian deities and cracking bas reliefs revealed little to the mystified French but the vengefulness of millennia of succeeding kingdoms and religions. Noses were hacked away (the Sphinx's nose, which legend states was chipped off by a Muslim fanatic in the fourteenth century, is only the most famous example of the wholesale desecration succeeding dynasties sometimes visited on their rivals), statuary was smashed, picture-script defaced. The damaged relics were mute testimony to how civilizations both destroy and build upon each other—literally. Pharaonic Egyptians built temples to Horus, Isis, and the rest of their pantheon; Greeks and Romans remade them in honor of their own deities; Coptic Christians defaced them and took pagan sarcophagi into churches to hold holy water; and Muslims salvaged ancient stones and columns for their mosques. Modern Egyptian peasants found more secular uses for the structures: they erected whole villages on top of some of the larger ancient buildings, drilling holes in the roofs as convenient refuse chutes.

Very few of the scholars were antiquarians, those quintessentially eighteenth-century characters, mostly wealthy, who filled their curiosity cabinets with strange old objects picked up on their travels, barely understanding what they had. Collecting old objects without understanding their use or meaning was a pastime for gentlemen, not a scientific undertaking. Only two of the original list of *savants* had any experience at all with artifacts.

Archaeology as we now know it was just beginning to emerge. In the previous century, Europeans had begun seriously investigating ancient sites, Pompeii in particular. Napoleon's scholars and engineers are remembered most as men who helped found archaeology as a science, their meticulous, systematic way of looking at ancient sites becoming the model for future Egyptologists. Unlike modern archaeologists, though, the French scholars did not excavate. First of all, they lacked the resources for excavation, and then, they found more to inspect and catalogue above ground than they had time for.

At first, the scholars only studied the antiquities in a haphazard fashion. They sketched or took notes on the relics they passed while searching for animals, rocks, and sources of water, sometimes chipping off bits and pieces of mummy and statuary, as countless visitors to Egypt had done since ancient times. Soon, however, the scholars grew so obsessed with the ancient artifacts that they eagerly braved furnace-like, Stygian interiors, aflutter with bats and musty with the bituminous stench of mummies, to record every inch of column, statue, and bas relief above the sand.

The scholars first began inspecting relics at Alexandria, where the mummies especially fascinated them. From those early days, the *savants* enthusiastically collected, dissected, and discussed mummies. They were also drawn to the colossal objects. At one early meeting of the Institute, Monge proposed sending a sarcophagus found in Alexandria back to France for further study. The sarcophagus did end up as part of the scholars' final collection but never made it back to France, as the English seized it. Others were content to study the relics *in situ*. Dolomieu's report on Pompey's Column correctly supposed that the hewn stone base was older than the column itself. An engineer presented the Institute with a paper on the Rosetta Stone shortly after that discovery.

Napoleon was never much interested in the antiquities beyond their potential as a symbolic backdrop for his campaign. He personally inspected only the pyramids and the Sphinx while in Egypt, shortly after arriving at Cairo, when he led some of the scientists and soldiers on an outing to the Great Pyramid. Although popular legend relates that the general experienced a mystical vision of Alexander the Great while inside the pharaoh's tomb, Devilliers's journal probably offers a more accurate account of what happened. According to his version,

Napoleon never entered the pyramid, the general being unwilling to demean himself by crawling on all fours through the narrow entryway.

Napoleon appreciated the emotional power of the ancient pyramids, obelisks, and temples, though without taking much interest in their real historical meaning. For the celebration of the anniversary of the French Revolution in early September 1799, Napoleon enlisted the scholars to help build wooden obelisks in Cairo with the names of the French dead inscribed on them, and ordered troops in Alexandria to hang the tricolor from the top of Pompey's Column. He later ordered French officers in Upper Egypt to celebrate the national anniversary on the ruins of Thebes, "the city of one hundred gates" as described by Homer, in the Iliad.

Probably more to stake a French claim in the sand than to found a new science, the last task Napoleon assigned his scholars before deserting them was to investigate the ancient ruins in the south of Egypt. Before this expedition left Cairo, however, one man was already well on his way to establishing himself as a field expert in Egyptian antiquities. A month after arriving in Egypt, Dominique-Vivant Denon attached himself to the military campaign chasing Murad Bey and his Mamelukes up the Nile. With the French troops, Denon spent nine months trudging a meandering trail along the Nile. He wove in and out of the desert, forded the river dozens of times, and reached the Nile's first cataract, on the edge of the region called Nubia (modern-day Sudan), before circling back toward Cairo again. He eventually published a two-volume journal of this voyage, which was translated into several languages and became a nineteenth-century best-seller.

On this journey, the elegant, lace-cuffed Denon proved himself a tireless intellectual trekker. He carried tomes by Descartes and Plato in his pack and always shivered in awe at his first glimpse of any ruins. Anxious to record everything he saw, he set up his easel and sketchpad in the burning sun while bullets flew around him. Some days he would ride far ahead, other days straggle behind. He was often out of his tent by dawn, drawing as fast as he could before the troops moved on. At least once he found himself in the middle of the fray, encouraging the French soldiers to return enemy fire by waving his drawing paper.

The soldiers marveled at the gray-headed artist's intrepidity and indulged Denon's crazy obsession with ancient Egypt because he was witty and charming and exceptionally close to the officer in charge of the campaign, General Louis-Charles-Antoine Desaix de Veygoux. Desaix, as he was known, was a cult figure almost as mythic as Napoleon. A year older than Napoleon, and by some accounts a full inch shorter, he died young, at thirty-one, his heroic legend unchallenged.

Desaix was notoriously rugged, and famously ugly. Napoleon called him "a little, black-looking man." The saber scar across one cheek did nothing to improve his looks. A Dutertre sketch shows a slope-faced man with a droopy mustache and receding chin. But Desaix's looks had nothing to do with what the man was about. Napoleon later recalled that "Desaix was wholly wrapped up in war and glory. He was always badly dressed, sometimes even ragged, and despising comfort and convenience. Wrapped up in a cloak, Desaix threw himself under a gun and slept as if he were in a palace."

Desaix was discreet and self-disciplined, though not so ascetic that he denied himself the small pleasures of a warrior. In Egypt, he assembled a seraglio of teenage slave girls. One fifteen-year-old Sarah, "a madcap Abyssinian," as he called her, given to him "as a present," became his constant companion on the campaign up and down the Nile.

Thus equipped, Desaix led repeated forays into the Sahara, tracking and battling the desert-tested Mamelukes. Murad Bey was a worthy foe for Desaix, more clever than fierce. "Sometimes he [Murad] pushed bravery to the point of madness, at other times he acted like a coward," wrote the Arab historian Jabarti. Murad's Mameluke ranks swelled with Arab fighters dressed in the Prophet's green who had sailed over the Red Sea from what is now Saudi Arabia and crossed the desert to help repel the infidel. These fanatics—the French called them "Meccans"—were answering a call for help Murad had sent to the religious leaders of Jedda, and they served themselves up as willing cannon fodder, astonishing witnesses like Denon with their sacrifice. Napoleon later wrote of them: "Their ferocity is equaled only by the misery of their standard of life, exposed as they are, day by day, to the hot sand, the burning sun, without water."

Accompanied by his own horsemen, the Arab jihadists, and armed gangs of peasants he had forced into service as he tore through their villages, Murad and the dregs of the Mamelukes led Desaix on a nine-month chase that exhausted the French soldiers, not to mention the region. The two armies skirmished with bloody results throughout the Nile valley, leaving terrified peasants and scorched villages in their wake. In the process, Murad also gave Denon—and, later, European armchair travelers—a fantastic tour of the ruins of ancient Egypt.

Riding through the fabled land of relics was just another magical episode in Denon's charmed life. As a seven-year-old, according to a story he later often retold, the boy aristocrat Dominique-Vivant Denon passed a gypsy girl on the roadside, her filthy hands outstretched. He reached into his pockets, pulled out a handful of coins, and dumped them into the little girl's palms. She smiled and shouted this prediction after him: "You will be loved by women; you will go to Court; a beautiful star will shine upon you."

By the time he reached Egypt, Denon had been loved by women, many women. He had gone to many courts. And a beautiful star, *une belle étoile*—Napoleon—was shining benevolently upon him. As a former gem keeper to the king, and then cataloguer of Napoleon's looted Italian art, Denon also had, more than most of the civilians in Egypt, the taste, training, and sensibility to appreciate what he was seeing.

Denon was a decided aesthete, but he traveled and supped like a soldier, suffering all the privations of the rank and file. He was covered with rashes and his eyes were usually afflicted with ophthalmia. He tried to treat them with honey and vinegar when available, but that was not often. "The surgeons, destitute of drugs, were only in the army to tell what ought to have been given," he noted.

As Murad Bey led Desaix and his men higher up the Nile, he frequently veered off into the desert, luring the French into barren desertscapes devoid of water or life. Food was scarce and the summer temperatures brutal. Soldiers wrapped their burning feet in linen, ate green dates, and stole the occasional peasant's goat to survive. Denon tried eating one of the vultures that always shadowed the army, but his Parisian palate couldn't be persuaded to finish the meal.

The artist soldiered on, fueled by his enthusiasm to see. He learned he could sleep almost anywhere. Exploring a temple at Thebes, he found a series of doors leading into ever gloomier rooms, until finally he arrived at an inner chamber he called "an asylum of terror." He would have drawn it, he wrote, but he was suddenly overcome by "so great a degree of lassitude, physical and moral," that he couldn't go on. When he emerged, the bright noonday sun dizzied him, the sand burned his feet through his shoes, and he leaned on a servant's arm, seeking shade. A cool, modern Arab tomb "appeared to him a delicious chamber," and he was just falling asleep on the cold stone inside when he was told that on a previous bivouac in the area, Arabs had slit a French soldier's throat on that very spot. Denon roused himself long enough to note the fresh blood smeared on the walls not far from where he lay, then fell back asleep. "The marks of this assassination filled him with horror, but he was so weary that he believes he should not have risen off the dead body itself of the unfortunate victim," he later wrote.

The greatest hardship of all for this creature of the salons was simply the bone-dry, pitiless *desert*—"that word terrible to him who has once seen the object it denotes!" he wrote. Desaix shared his feelings on the subject. At one point, Denon wrote, the general wandered past as the artist sketched a particularly barren vista near Thebes. "My friend, is this not an error of nature?" Desaix inquired. "Nothing receives life; all seems to exist for the purpose of saddening and terrifying."

Denon passed through Thebes and Karnak seven times on the fly with Desaix, never able to fully satisfy his need to see and draw his fill at those two spectacular sites. He lamented that it would take days simply to explore all the corridors in Thebes alone. In his memoir, he begged readers to forgive him for missing details. "Seated at his table with his map before him, the unpitying reader says to the poor traveler, 'Had you not a horse to carry you, an army to protect you?' 'Certainly, but have the goodness, reader, to recollect that we were surrounded by Arabs and Mamluks, and that most probably they would have seized, robbed, and killed me if I had ventured to go a hundred paces from the army.'"

Denon was a patriot, but his artist's soul recoiled at the cruelties of war. Between them, the French army and the Mameluke cavalry left destruction and terror in their wake. They demolished whole villages and burned crops, inflicting misery on the native populations. When he turned his infected eyes away from the antiquities, Denon witnessed the pillage with horror. On one desert trek, he found the remains of some of the Mamelukes' abandoned followers, mostly women, probably slaves or servants. Gazing at the corpses shriveling in the sun, Denon ruminated on the brutality of nomadic desert life. "The mind may paint to itself the lot of a wretch panting with fatigue and thirst, his throat shriveled, scarcely breathing the sultry atmosphere which devours him. The caravan is gone, already it appears to him no more than a moving line in space; soon it is no more than a point, and this point vanishes; he dies and his corpse, devoured by the aridity of the soil, will soon be reduced to a few white bones."

The French army left its own victims behind. The villagers and peasants in Upper Egypt were more terrified of French cannon than Mameluke swords and usually fled helter-skelter in advance of their arrival. A community of Nubians lived on the island of Philae, site of a great temple of Isis. The villagers were so afraid of the French that they plunged en masse into the river, drowning their smallest children rather than leave them to the tortures they expected the French would inflict. Here Denon found a little girl of seven or eight years, a recent victim of crude female circumcision. He wrote that "a cut, inflicted with equal brutality and cruelty, had deprived [her] of the means of satisfying the most pressing want [apparently, to urinate], and occasioned the most horrible convulsions; it was only by means of a counter operation, and the use of the bath, that he [Denon] was able to save the life of this unfortunate little creature, who was extremely pretty."

After rescuing the young girl at Philae, Denon looked around him and recognized the historic Isis temple, one of the largest in Egypt. Walking through its entry gates, he stood in the middle of acres of yellow stone courtyards surrounded by towering walls with mostly intact bas-reliefs. This breathtaking sight, known as the Temple of Philae, he wrote, made the awful day also "the most fortunate" of the entire campaign for him.

Denon surveyed the ancient relics of Egypt with the eyes of an aesthete schooled to revere classical Western art. At first, he wrote, the colossal structures inspired awe, though not admiration. He thought the sheer bulk of the monuments reflected a barbaric form of government. "The pride of raising colossuses was the first thought of opulence: it was not yet known that a cameo may be preferable to a colossal statue," he wrote. Denon later took a more favorable view, finding in ancient Egyptian architecture a "simplicity that produces the great."

Denon usually had no idea what he was looking at, no understanding of the real age of the sand-obscured structures or their original purpose, yet he often tried to guess. He couldn't read the hieroglyphic script. His drawings, though, were highly exact and crammed with detail, and he recorded page after page of his aesthetic impressions as well.

In nine months with Desaix, Denon made more than two hundred sketches, with precise architectural details and panoramas of the ruins and the desert. Denon spurned opportunities to quit Desaix's brutal march, and returned to Cairo only when the general did, in July 1799. He did not, however, decline the offer to sail home to France with Napoleon. As the first of the scholars to return to Cairo from Upper Egypt, Denon's stories whetted the appetite of every scholar within earshot. Napoleon spirited him back to Paris before he could read a full report to the Institute, but *Le Courrier* reported on what he'd seen, and published some of his drawings.

What Denon had started, the young scholars were already taking up in earnest. A group of them had been exploring Upper Egypt since shortly before the Syrian campaign, when Caffarelli had dispatched them to inspect the agriculture and irrigation systems on the upper Nile. Jollois, Dubois-Aymé, and Devilliers were among those assigned to do the dirty work of assessing the region for agricultural possibilities, a task that started with measuring the volume and rate of flow of the Nile at various points between Cairo and Aswan. But the engineers were ordered to camp in Asyut for two months while Desaix fought with Mamelukes farther south. The young men soon grew bored with life on the sleepy, sparsely populated banks of the Nile. Their commander, a senior military engineer named Simon LePère Girard, tried to keep the young engineers busy with water surveys, but after weeks encamped at the same spot, he'd exhausted all meaningful tasks.

The young men began venturing off on their own, exploring the Coptic ruins and tombs in the nearby hills. Girard was annoyed at this unauthorized fieldwork and tried to stop it. The young scholars posed enough trouble as it was, needing to be fed and minded, without sneaking off into unguarded areas, but the young men itched to explore, and they disdained their dry, technocratic commander, whose lack of imagination and curiosity irked them. Girard, who contributed a massive treatise on commerce, trade, and agriculture to the final book on Egypt, failed to comprehend why the ancient ruins so fascinated the young scholars. He complained often about all the "hieroglyph-making" by his young engineers.

At the end of May, Desaix allowed the engineers to move farther south with a military escort. Devilliers was in the throes of a nasty bout of ophthalmia, and he rode blindfolded, his seeping eyes wrapped in strips of linen, his horse dutifully following Dubois-Aymé's. After a short march, the group met Denon in the modern village of Qena, where he was bivouacked briefly with Desaix. The older man told them about a nearby ancient site called Dendara and a temple to Hathor, the cow-goddess, who ruled over music and lovemaking. He showed the young men some of his drawings of the temple, and told them he was especially excited about an intact interior ceiling bas-relief of the zodiac. He predicted it contained clues to the age and meaning of the ancient site, perhaps to the age of ancient Egypt itself.

Denon's report piqued the young men's curiosity and they snuck away to the temple at the first opportunity. The temple roof was covered with the ruins of brick houses, but its façade was pristine. Six perfect columns embedded with cow-eared Hathor-heads decorated the entrance and intricate bas-relief filled the shadowy interior. The young men were thrilled, but when they returned from the secret trip, Girard was furious. He and Dubois-Aymé fell into a violent quarrel, and Girard exiled the young hothead to the bleak outpost of Kosseir, across the desert on the Red Sea. Dubois-Aymé, as it happened, managed to profit from his punishment. He camped with Bedouins in the desert, later reporting in depth, for the scholars' final work, on nomadic customs and the Bedouins' generous hospitality to strangers. He also worked out an elaborate theory about the geological science behind Moses' Red Sea parting and other biblical stories that had the Red Sea and adjacent desert as backdrop. Two months later, Girard let Dubois-Aymé rejoin his friends on the Nile.

Devilliers and Jollois, more diplomatic, managed to get approval for their work in the ruins by taking their case directly to the general in charge of their military escort, Auguste-Daniel Belliard, who had a more intellectual cast of mind and gave them his blessing—and some guards. The two young men visited Dendara daily after that, making secret excursions into the shadowy confines of the temple, which was accessible only via a tiny window. Wriggling inside, they dropped down and stumbled over several corpses—fresh and mummified—in the dark. Apparently the room was sometimes used as a tomb for executed criminals, as the hangman's noose was still attached to some of the necks. Creeping around in the airless confines, feeling their way along guano-covered walls from chamber to chamber, their torches finally illuminated the zodiac ceiling Denon had described, an elaborate sculpted wheel beside a sculpted nude goddess, her arms raised skyward.

The young men decided Denon was right and that the Dendara zodiac was highly significant. For the next several weeks they went daily to the suffocating, black room, craning their necks up at the darkened ceiling, drawing until their torches burned down.

The ultrarational *savants*, of course, put no stock in astrology, but they believed the zodiac indicated where certain constellations were in relation to each other and to the earth, and therefore held clues about the age of the monument and the ancient Egyptians' degree of astronomical sophistication. They recognized certain symbols—the bull, the crab—and thought the zodiac proved that ancient Egyptian mythology was the basis for Greek and Roman mythology, and thus for the religious systems of the Western world. The zodiac did hold a clue to the age of the monument, but the French wouldn't realize it until decades later, and it was not what they expected. The Dendara zodiac was actually a more modern artifact, crafted not in the pharaonic era, but during the Greek era of Egyptian history.

At Dendara and later, the young engineers lived by one obsessive motto: *"Mesurer et dessiner"*—measure and draw. They were so fixated on the work that they weren't even deterred by a shortage of pencils. They made do with lead bullets left in Desaix's wake, which they melted and cast through reeds.

In his journal, Devilliers described the work of copying the zodiac as "long and arduous." Unlike Denon, who sketched on the fly, sometimes literally on horseback, the engineers approached their work systematically, applying drafting skills learned at their Paris school. Using string, they divided the intricate ceiling frieze into eight sectors and worked on a scale of one to five, reproducing every inch of it.

By June, the engineers moved south again, mostly at night in order to avoid the sun. The night journeys had a profound effect on the men, who were struck dumb by the combination of silver moonlight on the desert, the black, whispering river, and the sudden, towering shadows of temples and columns. "Night rides always have a grave and portentous quality that predisposes the mind to profound impressions," wrote the engineer Michel-Ange Lancret of his first sight of the Philae temple. "I reflected with a mingling of excitement, pleasure, and apprehensiveness that I was in one of the most extraordinary locations on earth, amid places that partake of the fabulous, the very names of which, recited since childhood, have assumed gigantic and almost magical significance."

The engineers and their military escorts reached the desolate edge of southern Egypt at Philae in early July. A harsh region at the best of times, Desaix's troops and the Mamelukes had depleted it further. The scholars subsisted on green dates and spent the midday hours hiding from the fierce light. After finishing their water surveys, the young men sketched the island temples of Elephantine and Philae. (These locations no longer exist as the French saw them. The Elephantine temple was destroyed in the decades after it was drawn by the French.) The French scholars were no more welcome than Denon or Desaix's army had been. The Nubians "threatened to tear down the remains of the temples if the scientists didn't leave," Devilliers wrote.

For hundreds of centuries, Philae had represented a major milestone for travelers in Egypt. To arrive at this place of narrowing rock walls and swift waters near the Nile's first cataract was to attain the southern reaches of the historic Egyptian nation, the very edge of inner Africa. The French, like tourists in Roman times, could not resist commemorating their arrival at this remote place on the temple's ancient stones, where they carved lists of their names. The island of Philae no longer exists. The Aswan High Dam inundated it in the twentieth century, and the Isis temple was moved, stone by stone, to another site. The *savants'* graffiti is visible today, high on a pale-yellow wall in a corner of one room of the vast temple, where tourists can still read the scholars' names, the date, and their precise geographical distance from Paris, calculated by the astronomer-priest Father Nouet. Home was, after all, always on their minds.

In late summer 1799, Denon's Institute colleagues set off to begin the archaeological fieldwork Napoleon had ordered. The *savants* fought amongst themselves about who would go and left in two groups. Fourier headed the first group of thirteen men, which included the musician Villoteau, Geoffroy Saint-Hilaire, and various artists, engineers, and architects. Mathematician Louis Costaz headed another group of fourteen men, including astronomers, architects, engineers, and naturalists Coquebert and Savigny.

The scholars sailed up the Nile in August 1799, just after Napoleon departed Cairo. They traveled on the double-masted boats called *dahabeahs* that still ply the Nile. Serendipitously, they left in August, when the Nile was at its flood height, and the north winds were strongest, and they sailed upstream at maximum speed. They reached Aswan in just four weeks, then circled back to Thebes, where they met up with the engineers in September.

The arriving scientists brought fresh tools and new hope to the young engineers, who had heard rumors of Napoleon's abandonment and felt forgotten. In one letter he wrote to Cairo just before the scholars arrived, Devilliers begged for more pencils and set the record straight on rumors of the engineers' demise. "We believed that you thought we were dead: I am letting you know officially today that we are doing very well."

The arriving scholars were astonished at the work the young engineers had already done on the ruins, and they agreed to work together to fill in what was unfinished. They descended upon Karnak and Thebes as a team, sometimes forming human chains. The young engineers focused on topography, for which they were trained, and surveyed the sites with plumb bob, compass, and T-square. Artists drew, astronomers calculated, and geologists examined and took samples. The job of copying the richly decorated walls and columns soon became a collaborative effort involving dozens of scholars. At Karnak, an assembly line of men copied every inch of the huge battle scene on one wall of a temple. The symbols were so numerous and time so short, the men often sketched them sloppily, such that they lost whatever significance they might have had for future translators.

Heaps of garbage and sand hindered them everywhere. At the Thebes temple, the debris and dirt had drifted up to the ceiling in some corners. Villagers employed the massive temple of Horus at Edfu as a garbage dump. Every morning peasants heaved buckets of ashes from cooking fires and donkey, camel, and horse manure through the window holes as the *savants* stood nearby with their sketchbooks and measuring devices. "One can easily conceive of the difficulties a European would encounter penetrating this subterranean fortress," Jomard noted dryly in his final report. (Edfu is now excavated and is the best-preserved major temple in Egypt.)

In their sketches, the artists and engineers drew the ancient sites as they would have looked new. They even occasionally inserted imaginary pharaonic priests into their drawings. Sometimes, the scholars drew themselves into their pictures and depicted the endeavor itself— a Mameluke-style tent erected near ruins for shade, small groups of Europeans in frock coats, sketching or measuring, vistas of sand in the background.

Archaeologists and plunderers have long since excavated these sand-drifted sites, which have been replaced with clean-swept, tourist-friendly parks. The *savants* left a record of sites buried in sand and rubble, as they had looked for centuries.

The scholars' work was thrilling and tedious. They tripped on fresh corpses and mummies and slipped in centuries of bat guano, in rooms so dark they couldn't see their own hands. Working by torchlight was dangerous in itself, since the long-enclosed areas were highly flammable, packed with wood, ancient paint, and mummy tar. They most feared not ancient spirits but stumbling into unseen holes, and the ever-present, fluttering *chauve-souris* (the French word for bat is literally "bald-mouse"), which swooped around their heads by the hundreds.

Jomard barely survived one bat attack. He and a companion were exploring a tomb at Thebes one late summer afternoon, each carrying a single lit candle. After following a mazelike trail deep inside, carefully skirting various plunging shafts, they faced a wall of intricate carvings. Stopped momentarily to admire it, hundreds of bats fluttered past, extinguishing their candles. In the sudden blackness, oxygen itself seemed to have been sucked out of the space. "Reason succumbed and imagination alone ruled," Jomard wrote. To fend off panic, they clung to each other's hands and inched along in utter silence, trying to avoid, by memory and feel, the chasms at their feet. Finally they saw a light in the distance—the exit. Reemerging in the setting sun, they were surprised to learn it was only six o'clock. The whole ordeal had lasted only an hour, but, Jomard wrote later, it felt like an eternity.

The scholars rarely knew what they were looking at: ruins of what they believed to be castles, temples, or tombs, or who or what a particular statue depicted or signified. They tried to apply scientific analysis, and where they could do so, they even debunked some legendary superstitions. The colossal pair of Memnon statues at Thebes were said to moan when the sun rose. The Greeks had a charming theory for it: when Eos, mother of the four winds, woke with the sun in the morning, her son, Memnon, sighed and groaned plaintively. The Roman emperor Hadrian himself had written that he and his wife heard unearthly sounds from the statue that "gripped them like nothing they had known before."

The scholars were having none of it. Devilliers wrote that he and his friends did indeed hear a "hollow creaking and cracking sound" just after sunrise, but they had identified a purely scientific cause. "The sound seemed to be coming from the enormous stones, some of which seemed on the verge of collapsing. This phenomenon probably comes from the abrupt change of temperature that happens at the rising of the sun."

Eighteenth-century science could put that ancient myth to rest, but it could not decipher the greatest riddle of all: the hieroglyphic symbols covering every inch of the ruins. Lacking the language key to the ancient world, to which these intricate scribbles spoke, they tried and failed to extract clues from picture images. One scholar erroneously wrote that the ancient Egyptians practiced cannibalism, basing his claim on a bas-relief of ritual military massacres. The *savants* also relied on Roman and Greek writers to tell them what the ruins had been. Using the classical writers' interpretations as a guide, for example, the scholars believed that buildings with non-military scenes on the walls were religious, and those with battle scenes were palaces. They did not know they were usually looking at temples, and that the scenes on the walls were entirely mythical, not historical.

The scholars also succumbed to their own supreme faith in mathematics. Jomard went to great trouble to measure the pyramids. Using a complicated theorem to compare the numbers he got (from his measurements) against the numerical latitude of Egypt, he divined that ancient Egyptians were so advanced they actually knew the circumference of the earth.

As misguided as their conclusions were, the scholars set the standard for modern archaeological fieldwork. The recently graduated engineers provided the rigor and energy behind the most detailed record yet of ancient Egypt for both armchair travelers and future scholars back home. But it was Denon, with his artist's eye and aesthetic sensibility, who first revealed to Europeans a deeply felt sense of Egypt, ancient and modern. His book, alive with awe, astonishment, reverence, was translated into Italian, English, Spanish, and German, and became the first best-seller of the nineteenth century.

CHAPTER 9

THE NATURALIST

Egypt, 1799–1800

A fulvous, arid nation bisected by a narrow strip of green fertility, Egypt teemed, surprisingly, with life. The hippos and gargantuan crocodiles floating down the Nile during the flood were only the most spectacular of Egypt's animals. The desert sand itself crawled with creatures Europeans had no names for. Even earthworms burrowed beneath the waterless crust. The marshes and riverbanks shrieked with birds, snakes nestled among the reeds in the mucky delta, and whole species of yet-unnamed fish and other aquatic life lived in the river, the salt lakes, and the sea.

After one desert trek from the Red Sea to the Nile, Denon described how the barrenness magnified life. "After a sojourn of eight days in the silence of the desert, the senses are awakened by the most trifling sensations; I cannot describe what I experienced when, in the night, lying near the banks of the Nile, I heard the wind shake the branches of the trees, and cool itself by filtering through the thin leaves; life was in the air, and nature seemed to breathe."

Dubois-Aymé, returning from his exile on the Red Sea, noticed the same thing. "When I arrived at the cultivated lands, I suddenly found Egypt beautiful, though I had formerly found it so sad. Although the palm-trees previously made me long for the forests of my homeland, after spending three months in the desert, these palm-trees and the Nile were full of freshness."

The ancient Egyptians observed this phenomenon, too, and their cosmology reflected a deep appreciation for the natural world. Mummified remains and bas-reliefs in the ancient tombs and temples attested to the ancients' reverence of the bee and sacred ibis, the apis bull, cats, saluki dogs, scarab beetles, scorpions, the asp on the crown of every representation of god and king. Temple walls were decorated with lion-headed, monkey-headed, hippo-headed, bird-headed, and crocodile-headed gods. Decorating temple and tomb, ducks and water plants were symbols of plenty, alongside the ubiquitous corn poppy and morning glory. Stone carvings of lotus and papyrus crowned colossal columns.

Napoleon's naturalists arrived in Egypt knowing very little about the region's flora and fauna beyond what the classical writers had said of the land and of the ancient civilization's reverence for indigenous animals. Only after a few months on the ground did they realize how much previously unclassified plant and animal life the seemingly barren land harbored.

In the eighteenth century, the term *naturalist* indicated a sort of general biologist who would today be classed into a subspecialty such as botany, ornithology, entomology, zoology, even geology—in sum, any of the fields that involve the study of life on our planet. There were a dozen "naturalists" on the expedition, including the teenage botanist Coquebert de Montbret, mineralogist François-Michel de Rozière, geologist Dolomieu, the botanist Alile Raffineau-Delile, the specimen painter Henri-Joseph Redouté, the botanist Savigny, and the mineralogist-turned-zoologist Geoffroy Saint-Hilaire.

The borders between their specialties were fluid. Most were as capable of studying birds, say, as flowers and fish. Because the naturalists on the expedition as a group were young, they were less confirmed in their fields of specialty when they left France, and were free to collect in Egypt that which most interested them. Twenty-one-year-old Savigny arrived in Egypt as a botanist with a special interest in birds. In Egypt, though, he collected thousands of invertebrates—insects and spineless sea creatures. Similarly, his slightly older colleague Geoffroy Saint-Hilaire studied to be a mineralogist but, thanks to the Revolution, had become a zoologist in Paris. In Egypt, he proved an expert ichthyologist. He morphed again, home in Paris, into a speculative, proto-evolutionary biologist.

Of all the civilians, the "naturalists" were the least useful to the military—indeed, by definition—and therefore the most subject to abuse from the soldiers. Because of his youth and relative high rank, Savigny felt the sting of military envy from the start. "I'm not very happy, we are piled on top of each other," he wrote to a colleague back in Paris from the ship. "The discomfort is aggravated by the hostility of the soldiers toward the scientists."

Their work was always hampered by the scarcity of willing military escorts (few soldiers were noble enough to think that guarding bird-watchers and bug-collectors constituted heroic service to France) and the lack of equipment. What's more, the naturalists lost most of their equipment before they even arrived at Cairo, when the ship carrying their scalpels, microscopes, tweezers, alcohol, jars, pins, pressing paper, and frames for mounting insects struck an offshore reef and sank.

The one advantage of their military uselessness, however, was that of all the *savants*, the naturalists were most able to devote their time in Egypt to collecting. In three years they filled thousands of jars and crates with specimens. Even at Alexandria and Rosetta, in the driest season after first arriving in Egypt, the botanists had collected seeds to send to the Jardin des Plantes.

It took Geoffroy Saint-Hilaire several months to create a lab from scratch in Cairo, but with his slave boy he soon was more than making do, scouring bazaars, fish markets, desert, riverbank, and seashore for new creatures. The naturalists were spoiled for choice, collecting, Geoffroy wrote, "with luxury." He amassed such a quantity of creatures, he could always choose the largest specimens, and sometimes a dozen individuals of the same species. He could barely keep up with the job of dissecting, preserving, and classifying everything according to the Linnaean system. He paid Turks and Egyptians to collect animals he or his slave didn't have time or energy to collect. Other scholars—and even the occasional friendly, science-minded military officer—routinely dropped new creatures on his and Savigny's doorsteps. Fourier brought them a "magic" snake, not because of its alleged immortality (villagers believed it was animated by the spirit of a long-dead sheik) but for its scientific significance. "We asked to see the spectacle, and we were shown it," Fourier wrote. "One of the people there passed the snake around our necks in order to pray to God that we be exempt from sicknesses and accidents. Before leaving, we bought the snake for 100 *medinas* because it was a type of snake that the naturalists had not described!" Neither Geoffroy Saint-Hilaire nor Fourier left any record of whether the mystical serpent was eventually immortalized in French taxidermy.

During their three years in Cairo, the naturalists collected thousands of specimens and made three major expeditions—to the Nile delta, Upper Egypt, and the Red Sea. In December 1799, Geoffroy Saint-Hilaire and Savigny, along with two astronomers helping to map Egypt and a few other members of the Commission, headed off on their first trip, into the Nile delta, a rich habitat for birds and fish. The tour began inauspiciously when the French commander of their boat refused to stop so they might collect specimens. They were finally rewarded when they reached the delta town of Damietta, located near the mouth of the Nile, and found the sea, the marshes, and the brackish Lake Menzalah teeming with millions of birds of every species. The wetlands of Egypt had been home to migrating birds since ancient times. Many of the same species that figured in the bas-reliefs on ancient temple ruins still flocked to the Nile delta, their habits unchanged over thousands of years.

Of the naturalists, Savigny was the most passionate about Egyptian birds. His journal is crammed with sketches and swooning entries on the copious avian life at Menzalah. "I have never seen such water birds," he wrote on December 10, 1799. "Flamingoes, cormorants, ducks, etc. At night, one can't hear anything but the calls of all these birds." A few days later, another entry: "Serene temperatures. Soft air. I have never seen so many waterbirds and so many species. The water's surface is rippling with wings. Multitude of herons. Indistinct cries everywhere. The effect is of gunshots."

Savigny drew dozens of these birds, in color and style rivaling Audubon, and contributed one of the most visually stunning sections to the *savants'* great book on Egypt. One of the birds he discovered was eventually named after him, *le guêpier Savigny*—the Savigny wasp-eater.

Savigny was so fascinated by the sacred ibis—the black and white bird that figures prominently in the old Egyptian religion—that he later wrote his own book on its natural history and place in ancient mythology. He put to rest the legend that the sacred ibis ate snakes, assuring readers that he had determined, by observation and dissection, that the bird ate only mollusks and shellfish. He proposed an anthropological basis for the ancient story. The ibis, he noted, always arrived in Egypt with the life-giving Nile flood, and thus became associated with positive human endeavors, including science and knowledge. Toth, the Egyptian god of writing and science, has an ibis head. The snake, as always, simply represented evil.

Savigny loved birds, but in Egypt he began collecting invertebrates, and soon spent most of his time poring over bees, snails, flies, spiders, mollusks, and worms. He worked constantly at his collections, but read just one paper at the Institute at Cairo, a study of the purple lotus. Unusually hardy and intrepid, he was rarely sick and was the only naturalist attached to the ill-fated Syrian campaign (Napoleon tapped him to go because Geoffroy Saint-Hilaire was temporarily blind from ophthalmia).

In Syria, Savigny filled hundreds of jars with insects and lizards he collected in the midst of plague and siege. Nothing survives of his observations, if any, from that dark saga. The only record of Savigny's work in Syria is in an upbeat letter Geoffroy Saint-Hilaire wrote to his museum colleagues later that year. Savigny, he reported, had "acted with great zealousness and disinterestedness in order to complete his objective" in Syria. Soaring personal expenses, disease, sandstorms, lost camels, bullets, cannonballs, and stolen horses did not deter him. "While the soldiers could barely survive the great fatigues of the desert crossing, Savigny spent his time collecting insects. The collection he made in Syria is very interesting, and he gave the lizards, snakes, and quadrupeds that he collected to me."

The naturalists soon found dead things in Egypt to be as fascinating as live creatures. They unwrapped and dissected the bituminous, shriveled remains of man and beast almost before they had unpacked their luggage in Egypt. On an early trip to the ruins of Memphis, near Cairo, Geoffroy Saint-Hilaire and Savigny descended into the darkness of the so-called Well of Birds, and found thousands of pots of mummified ibises lining the walls of the subterranean cellars. The pots were arranged horizontally, reminding the *savants* of bottles of wine in the caves of France.

Geoffroy Saint-Hilaire's colleagues back in Paris were awed by the mummy collection he brought home, regarding it as more impressive than his bottled and stuffed creatures or his new theories. In a report on Geoffroy's collection published by the Museum, Cuvier and others wrote, "One cannot master the transports of one's imagination when one sees again, conserved down to the slightest bones, least bit of hair, and perfectly recognizable, an animal that had, two or three thousand years ago, in Thebes or in Memphis, priest and altars."

Early excursions into the ruins around Cairo gave Savigny and

Geoffroy Saint-Hilaire a taste for archaeology, and both men were assigned to voyage to Upper Egypt. The convening of all the *savants* to work in the ruins was one of the happiest and most productive periods for all concerned, naturalists included. "That which took the *savants* several months seems as though it would have taken years," wrote Geoffroy's son, Isidore. "The most perfect harmony reigned among the members of the two Commissions and all were full of this ardor."

The naturalists made a third and final major voyage within Egypt to the Red Sea at Suez in late fall 1799. Napoleon had sent the engineers to survey the canal without naturalists, so Kléber, who organized this last mission, included almost all the naturalists—among them Savigny, Geoffroy Saint-Hilaire, Delile, Rozière, and a number of engineers.

Reaching the Red Sea was the high point of Savigny's Egyptian experience. "The findings he made surpassed even his dreams," wrote his biographer. At Suez he collected hundreds of tiny marine animals— starfish, sea urchins, brightly colored shells, coral, fish, crustaceans— which he painstakingly and beautifully later rendered in sketches. His drawing of a sea urchin collected on this trip is typical of his obsession with detail: a radiation of tiny needles, each of hundreds of individual spines firmly etched, it pulsates off the page like a spiralling hypnotist's eye.

In his unpublished notes, Savigny complained that his three days at the Red Sea were far too short for him to find what he sought: living polyps in the water. Unfortunately polyps, and all other specimens, large and small, had become insignificant as dire military events forced the naturalists back to Cairo.

In early winter 1800, Kléber negotiated his short-lived peace with the British and Turks, in which the French would retire honorably from Egypt, with their arms. The *savants* were promised safe passage home. Thrilled, the scholars rushed back to Cairo from wherever they happened to be in Egypt, carefully packed their specimens, drawings, notes, and souvenirs, and made for the coast.

At Alexandria, they boarded a ship, *L'Oiseau*—French for "bird"—but unfortunately this bird never flew. It floated in the Alexandria harbor for a month, the increasingly annoyed *savants* crammed together on board, waiting for authority to depart. After Kléber won decisively at Heliopolis, they disembarked, reboarded a *dahabeah*, and disconsolately sailed up the Nile and back to Cairo. In late summer 1800, the members reconvened the suspended meetings of the Institute of Egypt.

Unlike his colleagues, Savigny's energy never flagged in Egypt. He collected animals and insects and pored over them painstakingly with a magnifying glass for hours at a time, seemingly impervious to fatigue or the endemic illnesses afflicting everyone around him. Ironically he would suffer much worse years later at home.

Back in France, seemingly unscathed, Savigny embarked on an ambitious project of comparative entomology that involved examining the tiny jaws of butterflies and caterpillars in an era before high-powered lenses. For more than a decade he obsessively pored over what he had collected in Egypt. He drew and redrew 1,500 views of the mouths of insects in his collection. His goal was to show that the jaws of sucking insects are a derivation of the jaws of grinding insects, part of an evolutionary approach that would anticipate Darwin.

In 1815, Savigny started suffering debilitating headaches; he found the slightest amount of light excruciatingly painful and began losing his sight. Soon he could no longer identify his own prolific work. By his mid-forties Savigny came to fear sunlight like death itself. As his mysterious malady worsened, he bandaged his eyes and wrapped his head in black netting and took to his bed. When the weak European sun was strongest, in summer, he donned a mask of steel under his black netting. He installed steel shutters and thick draperies to extinguish dawn at his windows, and even in that contrived night, his mind's eye saw the shapes of what he first inspected in Egypt: the papery wing of a fly, tentacles of coral, the earlike curve of mollusk shell, the jointed legs of spider and scorpion, the infinitesimal hairs on the lip of the millipede.

He survived, nearly blind, for many more years, consumed with documenting the minutiae of his own bizarre ailment. In 1838, he submitted his last scientific paper—on himself—to his peers. "While common to all sorts of organs of the senses, this condition had its principal seat in the organ of sight. It could not, whatever its violence might be, bring about blindness, in the rigorous sense of the word, but little by little it rendered my eyes incapable of enduring light and in the ever more profound darkness into which it forced me to stay, it made a crowd of brightly colored images shine, of which the successive emissions, reiterated to infinity, tired me, haunted me incessantly."

Savigny and his doctors suspected he had contracted his eye troubles in Egypt, perhaps in the ruins themselves. "He brought back from [his] stay in Upper Egypt material of a first order, the study of which would absorb him until the crisis that would forbid all work to him," wrote his biographer. "It was there in the course of his research, in a desert country, in a burning climate, that he caught the germs of the sickness that would have such dire results for his organism."

Modern ophthalmologists have speculated that Savigny's illness was not ocular at all, but a form of epilepsy perhaps caused by a brain tumor, although they are at a loss to explain how he could have survived thirty years with such a condition. It's also possible his ailment was purely psychological. Savigny was willing to accept that his condition might be more emotional or spiritual than physical. He wrote that his pain and hallucinations stemmed from a "feeling of an invincible terror."

Savigny, a man of science, kept copious journals on his own condition to the last days of his life. He never knew it, but his work on insects advanced by a small step the still-inchoate theory of evolution.

CHAPTER 10

THE ZOOLOGIST

> *My dear Cuvier: Prepare yourself to make me the greatest sac-*
> *rifice. I demand nothing less than the anatomical throne. If you*
> *hesitate, I will reply to you: have you found in a single fish the*
> *organization of the quadrupeds and of the cuttlefish? Have you*
> *explained how the so-admirable organs of tetrodons act?*
> —Geoffroy Saint-Hilaire to Cuvier,
> writing from Cairo, August 16, 1799

By the end of 1799, the scholars were eager to leave Egypt. They knew Napoleon had authorized them to return to France at the end of that year, if they completed their work, and most believed they had done so. The generals, though, had other plans for them.

After Kléber was assassinated, in June 1800, General Jacques "Abdullah" Menou succeeded him as commander. If Kléber had been dealt a bad hand, Menou was left with plainly losing cards, but he refused to notice. Menou had converted to Islam and taken a Muslim wife (the daughter of a Rosetta bathhouse keeper) shortly after arriving in Egypt and was in no hurry to get back to France. He liked the desert climate and believed passionately that the French could colonize Egypt. He eventually proclaimed he would be buried in the ruins of Alexandria rather than give up Egypt, a vow he broke only after thousands of French perished under his command. Most of the army despised him for his devotion to the lost cause. They called him "Abdullah the Renegade." The engineers resented Menou for assigning them to public-works projects in remote provincial towns. The entire corps of scholars loathed him in the end for his officious personality.

Ordered to return to Cairo after Kléber's treaty of surrender collapsed, but unable to move back into their luxury quarters because of a raging insurgency in the city, the *savants* were temporarily homeless and miserable. A group of naturalists camped in a bleak fort outside Cairo, surrounded by the muck of Nile inundation. Four successive bouts of ophthalmia blinded Geoffroy Saint-Hilaire for a month, and dysentery left him leaf-thin and barely able to eat. "The whole world is attacked with little fevers and skin eruptions," he wrote Cuvier from the temporary hovel, adding that he, Savigny, and Delile had dysentery, too. "My birds are being eaten by insects, my collections are suffering. My cases of skins are destroyed."

When Geoffroy finally returned to Cairo, he bought his slave back from a local sheik and began to collect and categorize again, even though he was not sure how his work would ever get back to Paris. He scoured the bazaars, tombs, fish markets for specimens, talking with everyone from snake charmer to chicken farmer, dissecting and storing everything. He was proud of his vast collection, but in his darker hours, Geoffroy Saint-Hilaire imagined he might spend the rest of his life in Egypt, or in a Turkish jail. "Am I destined, after having begun a beautiful career, to live as a mercenary in the penal colony of Constantinople?" he wrote Cuvier. "I have made very precious natural history collections. . . . But there are so many difficulties to overcome! I admit, my good friend, that I am beginning to believe them insurmountable; I am resigning myself to share the army's fate; either the *savants* and the artists will remain with the army, buried in the sands of Arabia . . . or they will return to France, thanks to the indulgent goodness of our enemies."

Homesickness and uncertainty lowered the zoologist's spirits, but his mind was burning like never before, reaching pitched states of intellectual ecstasy that we, with the hindsight of modern psychology, might call mania. The day after Napoleon abandoned Cairo, Geoffroy wrote a letter to Cuvier in which he never even mentioned the general's drastic change of plans. He was more concerned with sharing some truly earth-shattering news: a new species of fish. The zoologist was actually more excited about sharing an *idea* he was developing from looking at the fish. This creature, a freshwater fish he named *hétérebranche*, possessed breathing organs with treelike bronchia similar to the human bronchial form. The similarity was another piece of evidence to bolster a theory of an *Ur*-form behind all life, a "unity of species," that was forming in the naturalist's mind. It was an idea so unorthodox even he would back away from it—for a few years, anyway—in the cooler latitudes of Europe.

Geoffroy Saint-Hilaire could never restrain his feelings for long, but his ardor was always at war with his ambition. Writing to Cuvier from Cairo on that hot August night, he quickly tempered his outburst of enthusiasm with apology. After his self-congratulatory paean, he continued: "It is far too much to boast of two dissections; let us talk seriously." He then described his latest work, adding, "I am so overwhelmed with business I do not know what I am doing or saying."

Geoffroy was also feeling giddy when he wrote that letter because he believed he would soon be allowed to go home. He predicted, in a letter to his father written the previous day, that he would be back in Paris in three or four months, with much of his work in Egypt already organized and ready for publication. That hope turned out to be overly optimistic. The zoologist and his peers in Egypt were not going to see France for another two years.

The worst still lay ahead.

Of the naturalists, Geoffroy Saint-Hilaire was the most prolific and wide-ranging. Besides collecting mummies and the living animals of the Nile river valley, he studied the ancient art of embalming and wrote reports on archaeological subjects. He presented numerous papers at the Institute, beginning with his investigations on whether ostriches could fly. Military observers who sat in on the meeting were outraged by the abstruse nature of the question, but his argument that the ostrich's vestigial wings proved that "nature never advances by rapid leaps" was also rudimentary Darwin. He read papers on the anatomy of the flamingo, the bat, and the crocodile, but his greatest output was on fish.

He was fascinated by the sharks, rays, and puffer fish of the Mediterranean, but also by other, previously unknown, freshwater creatures, including the lungfish he named *Polypterus bichir* ("this fish of which the discovery would have alone been worth the trip to Egypt," Cuvier later observed), and the flat fish with two eyes on one side of its head he named *Achire barbu*. In the Nile delta, the brackish waters were an inexhaustible source of new species and interesting observations. In Lower Egypt, where desert and fertile lands and fresh and seawater converged, he began to contemplate ideas that would occupy him for the rest of his life.

With only the younger Savigny familiar with zoology on the trip, Geoffroy Saint-Hilaire took liberties in his Egyptian work that he might not have taken back in France, under the critical eyes of his elder, more narrow colleagues. He turned his attention to thoughts beyond Egypt and began working in broad generalities—speculative biology—rather than specifics, an unacceptable practice in Cuvier's laboratories. "Further into the expedition," the science historian Toby Appel has written, "Geoffroy Saint-Hilaire lost all cognizance of the requirements of his colleagues at home."

The zookeeper of the Jardin des Plantes eventually arrived at a whole new way of looking at biology, which he called "philosophical anatomy." In Egypt, he began formulating a grand, new proposition, one that connected all life forms and involved invisible fluids and forces. Stripped of its more bizarre aspects, Geoffroy's philosphical anatomy anticipated the theory of evolution, by proposing that a Creator had *not* made each individual creature individually but had laid down a single pattern from which all living creatures developed. "It is known that nature works constantly with the same materials," he wrote in his fullest exposition of the idea. "She is ingenious to vary only the forms. As if, in fact, she were restricted to the same primitive ideas, one sees her tend always to cause the same elements to reappear, in the same number, in the same circumstances, with the same connections." While the fluids and forces he intuited were debunked, the idea that life forms had changed over time was a significant step in the direction of evolutionary theory.

In the fall of 1800, the French army was bedraggled, hungry, and going unpaid. Arabs were in near constant revolt; Turks were confronting French troops on Egyptian soil; and the British had locked down the coasts and were—unbeknownst to the French—preparing for a land assault. Tired of war and having exhausted their curiosity, the *savants* were ready to go home. Devilliers and Jollois planned to write up their reports on the ruins, acquire passports, and return to France. Menou, though, had other plans for them. By October, in furtherance of his delusion of crafting a successful colonial government, he ordered the engineers to report for duty in far-flung corners of Egypt. Most of the scholars thought Menou's dream was ridiculous, but the young men's energy and skills were the backbone of Menou's plan. By their last year in Egypt, the once-tender students had become desert foxes, leading their elders on marches into the burning wilderness of sand because they now knew where the wells were. Devilliers was teased throughout his life for one incident in the Sinai, in which he charged ahead with his guns drawn to meet the (fortunately friendly) Arabs on the horizon during a visit to the sulphurous Fountain of Moses. On one march back to Cairo from Suez, the same group ran out of water, and, searching in wells, found a two-meter-long snake instead of water. In a rage, Devilliers sliced it into pieces with a magnificent Mameluke saber a soldier had collected on the battlefield. The sword, he recounted later in his journal, "remained foul for a long time afterwards."

While the scholars plotted to get home, the soldiers took out their rage at being abandoned in Egypt by accusing the civilians of provoking the ill-fated invasion in the first place, and of hoarding looted treasure. They aired their suspicions and rancor at every opportunity. "Many soldiers called us useless mouths," Devilliers wrote. "Some of them believe that it was for the *savants* that they had fought the campaign. All the soldiers believe our boxes of antiquities contain treasures!"

In September of 1800, the Institute at Cairo reconvened its regular meetings after a long hiatus caused by Kléber's temporary surrender, a five-week Cairo insurrection, and the *savants'* own voyages. That fall, Geoffroy Saint-Hilaire read the first of a series of esoteric papers addressing general anatomical questions that had nothing to do with Egyptian animals. Under the cumbersome title "Exploration of a Plan of Experiments to Arrive at the Proof of the Coexistence of the Sexes in the Germs of All Animals," Geoffroy argued that he could demonstrate that the sex of an embryo was not predetermined, but influenced by its environment.

To prove his theory, he proposed experiments, at Cairo, involving at least 600 fertilized eggs and an incubator. There is no record that he ever received these eggs, but one can easily imagine the look on the face of the army supply officer asked to sacrifice food to such an experiment. During this period, Geoffroy Saint-Hilaire also suggested more elaborate experiments impossible to carry out in Egypt, and read papers on such matters as muscle contraction, without tying his theories to any specimens on hand. He also began speculating on the nature of dreams and sleepwalking.

His hypotheses and investigations never cured his homesickness, however. In fact, Geoffroy Saint-Hilaire's obsession with escaping Egypt grew. At the sixtieth meeting of the Institute, on February 25,

1801, Geoffroy proposed that the *savants* who had completed their work be allowed to return to France. The members decided they were not authorized to send their own members home, and the proposal failed.

Spring 1801 was the beginning of the end of the French campaign in Egypt. The British finally dispatched land forces to Egypt, and they landed at Abukir on March 8, 1801. This change in British tactics, from sea to ground, delivered the final blow to the French expedition. Menou moved himself and soldiers from Cairo to Alexandria and vainly confronted the fresh British troops, only to lose 4,000 French soldiers at the town of Canopus. Meanwhile, the Turks advanced on Cairo, where the French decided to surrender rather than fight. They had no choice: plague had entered Cairo, and was killing fifty soldiers a day.

On March 22, 1801, the Institute held its sixty-second and final meeting in the old harem quarters in Cairo. Geoffroy Saint-Hilaire read a paper on the anatomy of the Nile crocodile. Unbeknownst to him, London satirists would soon draw the French zoologist as a crocodile tamer with a bullwhip. Geoffroy's son and biographer would later try to set the record straight.

"It is false that he tried to tame the crocodiles whose habits he was studying at this time," Isidore pleaded, dead-earnest, in a footnote to his father's biography. "The officers of the English cruising fleet one day thought of an idea to amuse themselves and to take revenge for the bothers of their long stay on the Egyptian coast, by making caricatures of the principal personalities of the French army. These caricatures were sent to and published in England. Geoffroy Saint-Hilaire's paper on the anatomy of the crocodile, that had just fallen into English hands and that he had to later redo, became the text of the joke made about him: he was represented taming a crocodile."

Members at that final meeting also listened to a somber necrology report from Desgenettes. The doctor's count was unusually relevant that day, since the plague was killing high and low: the leader of the

Mamelukes, Murad Bey, who had survived the entire French occupation, was among the thousands to die of the plague during this epidemic.

The scholars spent their last days in Cairo gripped by panic and dread. With plague now in their midst, they were no longer as casual about the threat as when the epidemic was a distant rumor. Carts heaped with bodies rattled through the streets, trailed by wailing mourners and the newly sick. In a journal entry dated April 1–15, Jollois recorded the unfolding disaster. "Osmanlis' [Turks'] arrival by desert was announced and confirmed. All things precious to the French were brought into the Citadel. The civil administrations left. The Commission members ordered to stay inside. The plague made such horrifying ravages that one day it was counted that 900 Egyptians died and 500 Frenchmen were taken ill. The streets of Cairo presented every day the most terrible spectacle; one only met convoys of the dead, accompanied, one could say, by the dying. Many who followed the convoys got the plague, as could be seen by the signs that they exhibited, including violent headaches and vomiting. Such was the picture that the city of Cairo presented when the members of the Commission, justly alarmed, made the resolution to escape from the troubles that the war and the plague announced to them."

By the time the military command ordered the *savants* to evacuate Cairo and go to Alexandria, it was already too late for some. The epidemic stymied even the most rudimentary preparations. "At first we thought to go by land and went to the camp of some Arabs south of Old Cairo to buy camels," Devilliers wrote in late March. "But this tribe was terribly hit by the epidemic and many of our group came back infected."

Among those who caught the germ in Cairo was nineteen-year-old Ernest Coquebert de Montbret, who, though trained as a botanist, had served as Institute librarian. Like his mentor, Caffarelli, the young man was destined never to see France again. Throughout his three years in

Egypt, he wrote faithfully to his parents. Some of his letters were later collected and published. They form a touching portrait of a young scholar on the cusp of manhood.

"Plant gathering is dangerous here," he wrote in one letter, adding that Egypt was an ideal country for anyone with "ambitions to the title Martyr to Natural History." Later he wrote that he couldn't wait to see what his family thought of the changes in him after two years away. "I have become more confident, braver, more persevering and full of sangfroid," he reported. In his last letter home, he lamented the petty bickering in the Institute, and his lack of friends, complaining that his colleagues had become "cold and egotistical and like monks whose too-close society creates vices."

After he got sick, Coquebert's housemates, including the engineer Étienne Lenoir (who lived to invent the first parabolic mirror for lighthouses, and whose name is among those memorialized on the Eiffel Tower) cared for him as long as they could. As the French military situation in Cairo deteriorated, Coquebert's friends loaded him on a donkey and transported him to a friendly Copt's house in a more fortified quarter. He died there on April 7, a day after most of his fellow scholars had left Cairo, a martyr, as he had predicted, to natural history.

"I see him again," Devilliers wrote in his journal some days after the fact, "leaving our quarter by donkey to go to a faithful man who has promised to care for him after our departure from Cairo. He made a sign to me from afar not to approach him, because he had the plague."

Devilliers was by then in mortal terror of plague, but before leaving the city he had ventured out into the Cairo streets to retrieve his papers and drawings from where he'd stashed them in the Citadel to keep them safe from the war. He found the man in charge of the Citadel storage room laid low with plague, too. He took his things anyway, terrified that by touching them he would contract the germ. "The fruit of my work was half of myself, and I could not resign myself to losing it." Devilliers returned home to learn his servant had just died of the disease.

The scholars fled Cairo by boat, pursued by plague. "I had brought a mattress and a cover with me to sleep on the boat deck," Devilliers wrote. "One of us, who was a man of letters, LeRouge, having not taken the same precaution, asked me at night to share both of them, to which I consented, all the more gladly because he seemed indisposed. In fact, he was attacked by the plague, of which he died upon arriving in Alexandria!"

Within two weeks, 550 more French and 1,500 Egyptians in Cairo were dead of the plague, and the Turks and English were at the city gates. Conté, Desgenettes, and the musicologist Villoteau were among the few Institute members left in the city. Under terms of the surrender agreement, the French gave up Cairo, but were allowed to keep their arms. Desgenettes oversaw the evacuation of the convalescents, many onto British ships. He and Conté then boarded a boat to France—arriving home months before the rest of the scholars who had fled to Alexandria.

En route to Alexandria, the scholars had more to fear from French soldiers than from the English or the Turks. After starting out on boats, they soon had to disembark and travel by foot, dragging their cases and collections with them, because the English now controlled the Egyptian waterways. As they approached the dusty burg of Rahmaniya, still under French control, the French commander in charge of the region balked at letting the scholars pass. After arguing with him, they realized the commander so loathed them he "would have gladly thrown into the Nile" all their specimens and notebooks, Devilliers wrote. The starved and embittered soldiers, seeing the heaps of trunks, were furious at the thought of these civilians leaving Egypt enriched at their expense.

"Some of the excited soldiers wanted to pillage our treasures, but found only some debris and rock from Mt. Sinai," Geoffroy wrote later to Desgenettes. "This made the soldiers more indignant and the *savants* had to stay up all night to protect themselves." Even so, during the night, Devilliers lost a box of minerals and antiquities.

A captain of Napoleon's camel unit, the Dromedary Corps, finally took pity on the sorry group and escorted the scholars outside the walls of Alexandria. There, Menou ordered them to camp in quarantine for five days, as one of their group—the unfortunate bedmate of Devilliers—had just died of plague.

When Menou finally gave the scholars permission to enter Alexandria, they found themselves in a city under siege. With the French surrounded by the English and Turks on all sides, even the *savants* were conscripted into the national guard and given arms. "A veritable famine" was under way, Geoffroy Saint-Hilaire's son wrote. "Half of the army was in the hospitals or convalescing, and there were not enough doctors or medicine for the sick." Meat was rationed to a few ounces a week and the bread so salty it was inedible. Bedouins entered the city and sold corn for its weight in gold. Larrey, the doctor in charge of the garrison, finally persuaded Menou "after a very violent scene," Devilliers wrote, to allow the army's horses to be killed and cooked for bouillon. Menou had to admit the horses were now useless to the cornered French. The native residents of the city were barely subsisting on rotten grain.

From Alexandria, the besieged French could see, on the horizon, ships sailing back to France, carrying their comrades who had surrendered in Cairo. Desgenettes and Conté were among those lucky passengers. Desgenettes, for his part, could see Alexandria dwindling in the distance, and his thoughts were with his colleagues. "We finally set sail, but our gazes remained for a long time on the ancient and famous land from which we were moving away, and from Alexandria, where we left our fellow citizens, our friends, our brothers," he wrote later.

In the barricaded city, Geoffroy Saint-Hilare set up a makeshift laboratory, and there, in a rush of inspiration and sweat, he scratched out the first expression of his unifying biological theory on sheets of flyspecked paper. The once-pudgy scholar was skeletal, his clothes ragged, his long hair filthy and unkempt. An occasional breeze ruffled the heat in his room, bringing inside the stench and dust of the overcrowded city streets, where the oaths of sick, hungry, dispirited soldiers and the moans of starving natives were only rivaled in intensity by the regular prayer calls.

The zoologist was, however, engrossed in something more diverting than the French army's plight. Dissecting an electric fish hauled up from the Mediterranean Sea, he had finally discovered the link he'd been intuiting all along. To his mind, it was nothing short of the key to life, a principle so gigantic it unified all the sciences. For the first time in months, his zeal for study returned. He didn't even stop to think about how ironic it was that this burst of enthusiasm should occur at the moment when the French expedition to Egypt was reaching its disastrous conclusion, 15,000 dead men and three years after it began in the same dusty, miserable city.

Oblivious to the suffering and the dire military situation, he wrote frenetically, page after page, trying to transfer the light in his head into words on paper, to preserve this new idea that would, he was sure, make his name back in Paris. In the electric fish, he had discerned what he called a universal law that linked all life to the "imponderable fluids" of electricity, light, heat, and the animal nervous system. It is perhaps no accident that this cataclysmic unifying notion occurred to Geoffroy in the middle of the siege of Alexandria, although he later wrote that the bombs, the local fires, the attacks of the besiegers, and the plaintive cries of the starving natives evaporated for him as the electric fish revealed the link between "all the phenomena of the material world."

The zoologist's colleagues grew anxious about his obsession during the last three weeks in Alexandria. "His friends worried about him because his health couldn't resist such an excess of intellectual fatigue much longer," his son, Isidore, wrote later. "He himself felt it, but giving into the enthusiasm of discovery, he refused to allow himself to rest. The dearest of his colleagues, [and] Larrey himself—in his double authority as friend and doctor—were unsuccessful in convincing him. But Fourier went, and he had scarcely pronounced a few words when Geoffroy Saint-Hilare left his dear retreat."

What bombs and hunger had failed to do, the mathematician accomplished. A few words uttered by the usually sarcastic Fourier (who made no secret of his disdain for Geoffroy's "imponderable fluids") snapped the zoologist out of his research reverie and back into the real world. Fourier informed him that the British were inside Alexandria at last, and that General Menou, capitulating, had just signed away not only all of Egypt, but the entire body of the scholars' work— notes, specimens, drawings, collections—as well.

CHAPTER 11

THE STONE

> I have just been informed that several among our collection-
> makers wish to follow their seeds, minerals, birds, butterflies,
> or reptiles wherever you choose to ship their crates. I do not
> know if they wish to have themselves stuffed for the purpose,
> but I can assure you that if the idea should appeal to them, I
> shall not prevent them.
> —General Menou, in his capitulation to the English

Alexandria, August 1801

The final flight of the French from Egypt and the extrica-
tion of the scholars from their Oriental *foyer des lumières* is
an even more unlikely story than the tale of their arrival at the mystery destination three years before. The strangest aspect is that the
pudgy naturalist Geoffroy Saint-Hilaire should emerge as the scientific
corps's greatest hero. Nothing in his past predicted it. Like most of his
civilian companions on the expedition, he was neither an adventurer
nor a warrior. He was a scientist who happened to have come of professional age during the French Revolution. One of fourteen children
born to a not-so-rich barrister in a town just south of Paris, Geoffroy
Saint-Hilaire had only attended medical school to please his parents.
When the Revolution closed the royal Faculté de médecine, Geoffroy
Saint-Hilaire was free to pursue his real interest, earth science. As a

young student during the Terror, he played a small but brave role in rescuing some imprisoned clerics who had been his professors, and who were certain to be massacred. Afterward, he was rendered so ill with what later his son called a nervous fever that he took to his bed and had to convalesce in the country.

During the last month in Egypt, General Menou made the scholars and their notes and specimens pawns in a losing game with the British. Besieged in Alexandria, Menou refused to give the scholars permission to leave Egypt. After some cajoling, he agreed to allow the scholars to board a ship, but demanded they leave their collections, their drawings, and their manuscripts with him, calling them "a sacred trust." The scholars energetically refused.

Menou then revised his terms and demanded only the engineers' maps and plans. He also ordered everyone to sign a promise to take no document concerning the political and military situation in Egypt back to France.

Menou personally coveted only one treasure in the *savants'* baggage—the Rosetta Stone. He had been the general in charge of Rosetta when the engineers dug it up, and by virtue of that, he considered the stone his personal property. As soon as they entered Alexandria, he took it away from the scholars and into his tent for safekeeping.

In July, the scholars boarded the ill-fated *L'Oiseau* once again, and again it did not set sail. Menou withheld final approval for departure and the *savants* spent thirty-five more days on this boat, bobbing in the now sickeningly familiar Abukir Bay in sight of Alexandria. During these tense days, Dubois-Aymé threw a punch at the ship's captain, who lodged a complaint with Menou. Menou had the young man arrested, and Dubois-Aymé languished for weeks in a fort on the shore, convinced he was going to be left behind in Egypt. Devilliers had himself rowed ashore to comfort him. "He was very sad and one day thought he had the plague," Devilliers recorded in his journal about one visit to his friend. "At the moment when he showed this fear of having the plague, I kissed him, and denied his sickness. This, he [later] told me, calmed him and reestablished his confidence and his health."

After more than a month on the ship, Geoffroy Saint-Hilaire and two other scholars rowed ashore to personally beg Menou to release their boat. The general received the three messengers with a smile, chatted amiably, and promised to let them go home. "With an appearance of sincerity," Menou said goodbye, Isidore wrote. As an added show of good faith, he asked Geoffroy Saint-Hilaire to convey a ring to Joséphine Bonaparte from him, and even gave the ring to Geoffroy at the end of the meeting.

As it turned out, Menou still nurtured the delusion that he could soon get back to the business of administering a colony, and he did not want the scholars to leave Egypt. In a letter, he secretly warned Fourier of his true feelings. "Good citizen, I did not indicate any discontent regarding your departure either to the army or the government but your departure in the actual circumstances appeared to me, and still appears to me, and will always appear to me, immoderate and ill-conceived. But the lively manner in which I have expressed myself on this subject is entirely for your own personal attention."

There is no record of Fourier sharing Menou's "lively" communication with any of the other scholars, but the group soon discovered the general's displeasure in a more visceral way. Four days after the meeting with Menou, L'Oiseau lifted anchor. Once again, it was stopped, this time by two British cannon shots, naval code to retreat or be fired upon. The savants were indignant. Menou hadn't notified the British that the scholars were being sent home. With echoes of the British shots still ringing in the air, a French military dinghy approached the scholars' boat, bearing a messenger from Menou. The French general was threatening to sink the Oiseau in fifteen minutes if it did not set sail despite the British warning. The scholars suddenly noticed a French frigate in the water nearby preparing its cannons.

Fourier now leapt into diplomatic mode and personally rowed out to plead with the British sailors. The commander of the fleet, Sir Sidney Smith, who had watched Turks behead Frenchmen at Acre, agreed to let the scholars go, but retained Fourier as a hostage. The standoff ended only when the English finally persuaded Menou to let the *Oiseau* dock again at Alexandria. This episode destroyed whatever remained of the scholars' civil relationship with their last Egyptian commander.

Around mid-August, the English squadron began bombarding the French forts, and on August 31 Menou capitulated. When the English entered Alexandria, they were appalled at the starved Egyptian natives and infected, swollen Turkish prisoners, eyes blinded by infection, just released from a French dungeon, crawling on their bellies because their hideously swollen legs wouldn't carry them.

In the surrender agreement, Menou did the *savants* no favors. The British were demanding the large-scale booty—Nectanebo's sarcophagus, a colossal fist, and the Rosetta Stone. Menou, in a fit of loser's pique, agreed to give the British not only the large items, but also all the scholars' notes, drawings, and specimens, too. The British were thrilled. A geologist traveling with the British, Edward Daniel Clarke, went with British soldiers to inspect the scholars' materials. "We found much more in their possession than was represented or imagined," Clarke enthused in a letter from Alexandria in September. "Pointers would not range for better game, than we have done for Statues, Sarcophagi, Maps, MSS, Drawings, Plans, Charts, Botany, Stuffed Birds, Animals, Dried Fishes & c. Savigny, who has been years in forming the beautiful collection of Natural history for the Republic, and which is the first thing of the kind in the world, is in despair. Therefore, we represented it to General Hutchinson that it would be the best plan to send him to England also, as the most proper person to take care of the collection, and to publish its description, if necessary."

Savigny and the other *savants*, Clarke wrote, declined the invitation to England. "They said, perhaps the going to England would be felt as a palliation, if they had not been four years absent from France."

The scholars again approached Menou as suppliants, begging him to save their work. The general's response was to write a sarcastic letter to the British, leaving the matter between the scholars and the British command. "I have just been informed that several among our collection-makers wish to follow their seeds, minerals, birds, butter-flies, or reptiles wherever you choose to ship their crates," he wrote to General John H. Hutchinson, the commander in charge of the siege of Alexandria. "I do not know if they wish to have themselves stuffed for the purpose, but I can assure you that if the idea should appeal to them, I shall not prevent them. I have authorized them to address themselves to you."

The scholars believed the British sought their intellectual harvest with a devious ulterior motive. They suspected that the diplomat and antiquarian William Hamilton, in Egypt on a mission for Lord Elgin (of Greek marbles fame), wanted to steal their hard-won knowledge and claim it for his own. A trio of naturalists—Geoffroy Saint-Hilaire, Savigny, and Delile—threw themselves on Hutchinson's mercy, in a personal mission to the enemy camp. Hutchinson received them "politely but coldly," Isidore wrote.

It was at this meeting that Geoffroy famously rose to the occasion, becoming a hero to French science for the ages. The zoologist made a passionate plea to the English, arguing that only the French scholars could possibly decipher their own notes, sketches, and specimens. Various flowery recitations of this incident are recorded in French memoirs of the war. The official tome *L'Histoire scientifique et militaire de l'expédition de l'Egypte* (The Scientific and Military History of the Egyptian Expedition) quotes Geoffroy Saint-Hilaire thus:

"You are taking from us our collections, our drawings, our plans, our copies of hieroglyphics, but who will give you the key to all of this? They are only preliminary drafts that our personal impressions, our observations, our memories must complete. Without us, these materials are a dead language, from which you will hear nothing, neither you, nor your *savants*. We have spent three years conquering, one by one, these riches, three years gathering them from all of the corners of Egypt, from Philae to Rosetta: to each of them is attached a peril surmounted, a monument seen and engraved in our memories. Now we find here, on this frontier, a camp of soldiers who have transformed themselves into a body of customs officers, to stop and confiscate these products of observation and of intelligence! It will never be so. Rather than allow this iniquitous despoliation and vandalism, we will destroy our property, we will disperse it in the Libyan sands, or we will throw it into the sea. Then we will protest in Europe, and tell by what violence we were reduced to destroy so many treasures!"

A tense standoff ensued, Hutchinson refusing to relent. He sent the scholar Hamilton to tell the *savants* to hand over their materials at once. Geoffroy reiterated to Hamilton that sooner than sacrifice their belongings to the British the scholars would burn them. He then asked Hamilton if he wanted to be remembered for the burning of a body of knowledge comparable in richness to the Ptolemies' library at Alexandria.

Hamilton either believed the threat or took pity on the belea-guered intellectuals—possibly both. He went back and pled their case with General Hutchinson, predicting that the enlightened part of the English nation would indeed think poorly of them if they seized the French *savants*' treasures for the English, or allowed them to destroy their finds.

The general relented and agreed to redefine the surrender terms to let the French keep their personal property.

When Hamilton delivered this happy news to the *savants*, he again proposed that they return with him to London to publish their findings under the friendly auspices of the British government. The scholars again declined the invitation.

Geoffroy Saint-Hilaire modestly never mentioned his heroism in letters home. Having made his great discovery about the unity of life, he now wanted nothing more than to get home and share it. Writing to Cuvier from Alexandria in late September 1801, he complained that he had nothing left to do in Egypt. "While I vegetate and lie dormant here, you advance science," he wrote. "In truth, nothing is more despairing to me than this prolongation of our stay. My collections are becoming ruined and I have already lost four collections of birds; those that are conserved in liqueur are also suffering a lot, the spirits of wine evaporate, and if my stay prolongs itself, I will only bring you back mummies."

At the end of September, the scholars finally began boarding a variety of ships for the trip back to France. At Marseille and Toulon they were clapped into quarantine for another month. It was 1802 before they were all finally back with their families and colleagues. The scholars returned to France with their notebooks and specimens mostly intact. The British had helped themselves, however, to the larger treasures, including, most famously, the Rosetta Stone.

Word of the stone's discovery had reached France and the rest of Europe while the scholars were still in Egypt. Long before it arrived, it provoked much interest among scholars. The scientific study of language had been increasing during the eighteenth century with the typological work of the French *encyclopédistes* and other efforts to discover whether all languages might not perhaps be rooted in a single, ancient language given by God to Adam. The study of the Egyptian hieroglyphs had been nearly impossible, though, because before the scholars went to Egypt and copied down everything they saw, no large body of hieroglyphic text was available in Europe. Furthermore, the Europeans lacked any knowledge at all of the sound of the ancient Egyptian language. They assumed that the hieroglyphs were symbols, not sounds. The stone, with its trilingual inscription, was destined to change all that.

Menou didn't give up the stone without a fight. He argued with the British that the tablet was his personal property, as much "as the linen of his wardrobe or his embroidered saddles." Hutchinson wouldn't hear of it. On September 12, the British sent two envoys to negotiate, geologist Clarke and William Hamilton. Clarke stationed himself outside Menou's tent while Hamilton went inside. He overheard Menou shouting indignantly. "The words *'Jamais on a autant pillé le monde!'* [Never has the world been so pillaged!] diverted us highly, as coming from a leader of plunder and devastation," Clarke wrote.

The British chroniclers might have been exaggerating a bit, but Menou's churlishness was documented to the end of the affair. In a letter to Hutchinson regarding the stone, he wrote, "You want it, *Monsieur le général*? You can have it, since you are the stronger of us two. . . . You may pick it up whenever you please."

A French officer and member of the French Institute eventually retrieved the stone from beneath some mats in a warehouse with Menou's baggage and handed it over to the English in the streets of Alexandria. Colonel Tomkyns Hilgrove Turner wrote later that he "carried off the stone, without any injury, but with some difficulty, from the narrow streets to my house, amid the sarcasm of numbers of French officers and men." He also reported that when the French learned the English planned to take the stone, "they tore off its protective packaging and threw it upon its face."

All was not lost, of course, since the French had already copied and cast the stone several times over. The first of those proofs had long since reached Paris in a general's baggage. The original Rosetta Stone traveled to England on HMS *L'Égyptienne*, accompanied by Turner, who described it as a "proud trophy of the arms of Britain—not plundered from defenseless inhabitants but honorably acquired by the fortune of war." The stone arrived in England in February 1802, and was placed in London's Society of Antiquaries, where copies were made and sent to universities in Oxford, Cambridge, Edinburgh, and Dublin. Experts immediately went to work trying to decode the writing. In June, King

George III officially donated the actual stone to the British Museum, where it resides to this day.

In losing the stone, the French forfeited their single greatest Egyptian find to their foremost enemy in the world. They never forgot it. Two decades later, a young French linguist, Jean-François Champollion, beat an English scholar, Thomas Young, to the decipherment of the hieroglyphic script, using copies of the stone as his guide. A number of the *savants* were still alive in 1822 when Champollion finally succeeded in translating the pictures into sounds that made recognizable words, including a king's name—Ptolemy—written in the Greek translation on the same stone. Champollion's discovery was not universally accepted. Jomard, for example, refused to believe the symbols were really sounds, insisting throughout his long life that they must represent some higher significance than mere turnings of the tongue.

Translated, the stone turned out to be a decree written in 196 B.C., at the very twilight of the ancient Egyptian era, from one of Egypt's late Greek rulers, Ptolemy V. It detailed various repealed taxes and instructed that statues be erected in various temples, and that the decree be published in the writing of the speech of the gods, the writing of the people (demotic), and the writing of the books—namely, Greek.

After nearly two millennia, the "writing of the speech of the gods," as the stone described the hieroglyphs, was suddenly intelligible to men. Champollion's feat transformed the study of Egypt from one of simply looking at mysterious ruins and guessing at their meaning, to one of real understanding of a great but vanished human civilization.

One of the greatest ironies of the entire Egyptian campaign involved the stone. The French who surrendered at Cairo in June were allowed to keep all their belongings. Had the scholars simply left the Stone behind in plague-stricken Cairo with Conté and Desgenettes, instead of hauling it to Alexandria with them, it would probably be in the Louvre today. Instead, it rests in the British Museum, where no footnote on its glass enclosure reveals to visitors the French role in its discovery. If you lean in close you'll see the tiny white painted inscription, which tells only part of the story: "Captured in Egypt by the British Army in 1801, Presented by King George III."

CHAPTER 12

THE BOOK

> By deploring the fate of so many brave warriors who have
> succumbed in Egypt, the existence of such a precious work
> will console. The time will come when the same army, now
> occupied sneering at our faces, will pride themselves on having
> seen them and having known us. Let us all wait, and let us
> know how to suffer here patiently.
>
> —Geoffroy Saint-Hilaire, in a letter
> to Cuvier from Egypt, 1801

France, 1801–1828

They limped back to France in late fall on board British and Greek merchant and naval vessels—no French ships—with names like the *Amico Sincero, Callipso,* and *La Diane.* General Menou, one of the last French to leave Egypt, had the double misfortune of capitulating and catching plague. He sailed back to France under care of Dr. Larrey, with his Egyptian wife and their infant son. He survived the disease, but unlike the gloriously dead heroes Kléber and Desaix, Menou is not commemorated with a street name, plaza, or statue in Paris.

Devilliers, Jollois, and Dubois-Aymé sailed home on a rickety boat manned by Greeks. Nearing Genoa one night a ferocious storm nearly shipwrecked them. Several sailors were lost overboard and masts cracked. Devilliers stripped off his clothes in anticipation of having to swim. The sun rose the next morning on calm seas and the young men thereafter had an uneventful trip.

Geoffroy Saint-Hilaire lingered impatiently in Egypt waiting for a berth, and finally boarded a ship at the beginning of October. He sailed home on an English boat, the *Callipso*, with a small group of French engineers, forty crates of specimens, and some live animals. His nonhuman traveling companions included an Egyptian mongoose ("tamed," he reported to the Jardin's stand-in zookeeper, "to an extraordinary point, its gayness is remarkable"), Menou's pet civet, and two Tunisian genets.

Of the 151 French civilians who had arrived in Egypt in 1798, twelve left before surrender, twenty-six died in Egypt, and five more died shortly after returning to Europe. The toll on the military was much more devastating. Of the 34,000 land troops who sailed to Egypt with Napoleon, 21,500 returned alive. Among the survivors were 3,000 sick and maimed. Out of the expedition's 16,000 sailors, only 1,866—barely one in nine—are known to have returned to France. More than that probably survived the Abukir Bay disaster, but they were absorbed into the army and returned to France as soldiers.

At Marseille, on French soil at last, the returning *savants* did not meet a hero's welcome. They were clapped into quarantine for a month and housed under guard near the port in a warren of wood-frame sheds and warehouses, some without roofs. They sheltered themselves as best they could in the early fall chill with bits of sail dragged over from the ships, sleeping on hammocks. Their specimens and souvenirs—packed in cotton—were quarantined even longer than the men. The French suspected that cotton transmitted *la peste*.

Geoffroy Saint-Hilaire arrived home physically healthy but emotionally precarious. He was deeply worried about his professional future, having received not a single letter from Cuvier or his Museum colleagues for three years. For the first time, in a letter dated November 18 from Marseille, Geoffroy conceded to Cuvier that he "should have listened to" colleagues and friends who told him not to go to Egypt in the first place. This fit of melancholy was soon replaced by a more immediate, material concern: Where would he live in Paris? The question of lodging was much on his mind in the quarantine sheds. Sometimes he wanted his old apartment back; sometimes he imagined he could be quite happy living in a Spartan, single room. Soon after arriving in France, Geoffroy learned that Dolomieu had died in Paris. He immediately wrote to Cuvier, expressing his shock and sorrow—and inquiring as to his chances of securing Dolomieu's Paris apartment for himself. A month later, he wrote to Cuvier, complaining that he'd been wronged because his Paris apartment had been taken from him while he was away. Typically, though, Geoffroy was of two minds about this, too. In the same letter, he wrote that he'd lived in "a rough tent" in Egypt for so long that when he returned to Paris he planned to rent a simple suite.

The question now lay before them: How to make sense of what they had collected? Morally, too, how to account for what had happened? Was it possible to atone for all those soldiers' lives?

The scholars had been out of quarantine and back with their families for just two months when Napoleon issued a consular decree in February 1802, authorizing the publication of "the memoirs, plans, drawings, and generally all the results relative to science and art obtained during the course of the Expedition." The scholars had been planning such a project themselves. Fourier had discussed with Kléber collecting the savants' work in a single book. As first conceived, the book was a private venture. A French merchant in Cairo formed a joint-stock company with fourteen of the scholars to eventually publish the work in France. Menou later objected to the stock company, provoking the scientists to begin worrying about who would control their work. Back in France, Napoleon's government disbanded the stock company immediately and put the project under the aegis of the Ministry of Interior, which agreed to pay the scholars for their work on the book.

Begun in 1802 and not definitively completed until 1828, *The Description of Egypt* is itself a monument, if not, as Geoffroy Saint-Hilaire hoped, a consolation for those killed or maimed in Egypt. The book is lush, gigantic, astonishing. Its twenty-four volumes comprised the most comprehensive view of the culture and architecture of ancient and modern Egypt ever seen in the early nineteenth century. Nothing like it, in fact, had ever been published, on any subject. Never before, and possibly never since, at least in the predigital age, had so much information from so many sources and in so many forms (text, drawings, maps) been compiled into one work.

The book was published serially between 1809 and 1828 and sent to wealthy subscribers bound in brown paper. Subscribers could also order an elaborate mahogany storage chest designed by Jomard and carved with motifs of lotus flowers and busts from Thebes and Dendara. The production of so many oversize pages employed whole streets of paper-makers in Paris. It was expensive to print and built to last, on heavy linen pages. Only one thousand copies of the first edition were printed.

The book took so long to finish that dozens of men involved in the expedition died or became incapacitated before it was published. Napoleon himself rose from consul to emperor, then fell to citizen-in-exile and died, but not before Louis XVIII, brother of the executed Louis XVI, replaced him as monarch. Seven years would pass after Napoleon's death before the last volume was published. One of the trickiest aspects of the job was how to reconcile the scientists' Napoleonic hero worship with the newly restored monarchy. Geoffroy Saint-Hilaire's son and biographer, Isidore, wrote that the monarchy's restoration considerably slowed work on the project. The Bourbons felt no great love for scholars so intimately associated with the citizen-emperor. Rather, they set out to punish a number of them. "The fall of the Empire almost caused its abandon, for the book was too grandiose to have any editor but the government itself," Isidore wrote. "The least that had to be done was to erase Napoleon's name from this monument, as from all the others."

The scholars scoured the text and illustrations aiming to erase or at least diminish references to the emperor. It was no easy task. For the initial volume, the Egyptian expedition's artists had designed an elaborate frontispiece glorifying their leader's triumphal entry into a mythical Egypt, conquering Mamelukes, and trailed by the muses of Science and the Arts. Some of the volumes were published before Napoleon's death, but he was dead years before the full project was finished. In 1828, the frontispiece was still there, but all visual indications of the emperor were absent.

Fourier, assigned to write the preface, walked the tightrope between royals and emperor by resorting to shameless patriotism, touting the glory of France. He needed to make the expedition look successful in civil terms, though it had been a disaster for the French military. He wanted to glorify his mentor, too, and some accounts suggest that Napoleon personally inserted lines into Fourier's writing.

Fourier has been described as the scholars' chief ideologue, because of the strongly political context in which he placed the expedition and their work in his preface. He did certainly put a lofty spin on the French goal in Egypt: "[T]o abolish the tyranny of the Mamelukes, to extend irrigation and cultivation, to open a constant communication between the Mediterranean and the Arabian Gulf, to form commercial establishments, to offer the Orient the useful example of European industry, finally to render the constitution of the inhabitants softer and to procure them all the advantages of a perfected civilization."

Rewarded by Napoleon in January 1802 with the prefecture of Isère in Grenoble, Fourier was also busy with other matters, including working out a groundbreaking theory of heat and the draining of Grenoble's malarial swamps. He complained that he couldn't write the preface until the work of all the tardy contributors was in front of him.

The tardiest of the contributors was Savigny, creator of hundreds of animal and insect drawings for the book, who still had not finished his work by 1815. Worse, he was visually incapacitated and could no longer identify his own drawings. The editors of *The Description of Egypt* made a point of explaining, at length, in the final text that Savigny was too ill to contribute explanations of his work, and that the job had been passed on to another, with hopes that someday Savigny—"a victim of his devotion to science"—would be able to write his own text. The editors even saw fit to include in the book their letter to Savigny informing him of their decision to assign a student, Victor Audoin, to finish his job.

In their *Description*, the scholars tried to create an Egyptian version of Diderot's encyclopedia. They adhered to principles of objective, rational thought, elevating mathematical precision over all. Beyond the beautiful engravings, the book is memorable for its wealth of numerical interpretation and maps. Some modern critics see this reduction of people and places to numbers and maps as a flying wedge of colonial technocracy. It can also be interpreted as the scholars' way of trying to order their own experience. Abandoned by Napoleon, cut off from friends at home, the scholars in Egypt were afflicted with uncertainty about their own futures as much as by the perceived chaos of the Egyptian way of life. Numbers and maps offered a kind of comfort under such conditions.

Art historians and archaeologists study the book now, and beyond that group, few people ever leaf through its oversized pages. Copies of the original are few and far between, stashed in major library vaults in the United States and Europe, and in private collections around the world. The panoramic scenes of ruins half buried in sand depict a bygone Egypt that is simultaneously nostalgic, visually appealing, and informative. Savigny's delicate color drawings of birds rival Audubon's, and his infinitely various, mesmerizing black-and-white drawings of thousands of tiny invertebrates are among the finest examples of nature drawing. The section on modern Egypt with Conté's and Dutertre's illustrations of living Egyptians in their workshops and at rest are a beautiful, sensitive, and accurate record of a vanished era in Cairo.

Whether the book was a "consolation" for the disastrous Egyptian campaign and the immense and pointless losses suffered by the French military, as Geoffroy Saint-Hilaire had once predicted it would be, is certainly questionable. At the least, the book set the record straight on what the scientists were trying to do there.

Like Rip Van Winkles, the scholars returned to France to find that Paris and the world had moved on without them. They had spent nearly four years of one of the most unpredictable, tumultuous eras in European history in a faraway desert land with little or no news from home. Friends and family members had married or died, colleagues had been promoted. Geoffroy arrived unsure whether his father was alive or not (he was). As for their careers, science had advanced without them, and without their knowledge. In 1799, the British scientist Humphry Davy helped show that heat was motion not "caloric." In 1800, a German scientist named Karl Friedrich Burdach suggested the term *biology* as a replacement for *natural history*, the traditional name of the field which comprised zoology, mineralogy, and botany. The same year, Alessandro Volta announced the first electric battery. In 1801, the English scientist Thomas Young (who also studied the Rosetta Stone) showed that light moved in waves. The scientists returned to France unaware of these and many other groundbreaking discoveries in their respective fields.

They also found France itself profoundly changed. Politically, the nation was transformed. Napoleon the young citizen-warrior was now First Consul of the French Republic; he had also reconciled with the Catholic Church and was well on his way to becoming emperor. This last development would be profitable for the scholars, because the emperor was not going to forget his encyclopedia.

The *savants* returned to France altered in a variety of ways, depending on age, temperament, personality, and constitution. They had grown up or lost their health, embarked on previously unimaginable projects, found new interests or refined previous ideas. Paris looked outwardly the same, but in many ways it was a different city from the one in which they had come of age. That Paris, skeptical to its core, was gone forever.

In the years between the *savants'* return from Egypt and the publication of their book, France and Europe entered a new era. The *savants'* lives straddled the end of the Age of Reason and the beginning of the Romantic era in European cultural history, and in some ways they embodied the transformation. While their book was a sometimes dull Enlightenment masterpiece of classification and enumeration, the Egyptian campaign itself was the ultimate Romantic adventure. To venture on dangerous, distant voyages and risk life and limb for the sake of Knowledge was the stuff of the novels and poetry of the era. A by-product of the Romantic era was the Gothic, which evinced a taste for all things exotic and mysterious, and the scholars did indeed bring back such things from Egypt in their sketchbooks and specimen jars. The *savants* were skeptics, but they lived Romantic lives. They had survived the Terror and wiped the blood off their shoes, literally or figuratively in Paris. Such men would—one might imagine—never have been fully satisfied with life inside a museum office. Like characters in the fiction of the age, they were eccentric, tragic, doomed.

Napoleon himself was the quintessential Romantic figure. The citizen-warrior rose from common soldier to proconsul, then emperor, ruled for almost fifteen years, suffered defeat at Waterloo in 1815 and was exiled, to be replaced by the old royals and clerics. Even before he left, the Age of Reason—of classification and strict skepticism—had passed and a new taste for mysticism replaced it.

In exile on St. Helena, he never tired of talking of Egypt, the period in his career when greatness had first begun to seem possible. On that lonely, windswept rock, the fallen emperor had the leisure to daydream about the desert, and the audience to lap up his revisions and improvements upon history. His courtiers, the loyal few who retreated to St. Helena with him, wrote down every word—they had nothing else to do. They encouraged the great leader with questions and reminders. In turn, he set the record straight. He wasn't alone in wanting to rewrite his own history, but he was more deluded on the topic than most.

"In Egypt," he told one of his scribes, the Count de Las Cases, "I found myself relieved of the check of a bothersome civilization. I dreamed of all sorts of things. I created a religion, I saw myself on the road to Asia, carried upon an elephant, a turban on my head and in my hand a new Koran that I would have composed to suit my needs. The time I spent in Egypt was the most beautiful in my life, because it was the most ideal."

Napoleon's memory of the Egyptian campaign as an Oriental idyll animated him till the end of his life. He always contended that all of Asia had been within his grasp, and would have been the key to a more enduring greatness. He denied that he had abandoned his men, or ordered plague victims poisoned. "My departure from Egypt was the result of a grand and magnanimous plan," he told his listeners in St. Helena. "How ridiculous is the imbecility of those who consider that departure as an escape or a desertion."

In his last days, he became ever more delusional on the subject. He informed his companions that the name Napoleon meant "lion of the desert" and that he felt he had been born to rule that place. "The desert always had a peculiar influence on my feelings," he told Las Cases. "I never crossed it without a strong emotion. It seemed to me the image of immensity. It had no boundaries, no beginning, no end. An ocean on terra firma."

Like Napoleon, most of his scientists spent the rest of their lives reliving, regretting, or just musing about their three years in Egypt. To the end, most of their fortunes were tied directly to the rise and fall of the general who had led them to the desert.

As he rose to prominence, Napoleon dispensed titles to his favorites. He made Monge and Berthollet counts.

Berthollet accepted the honor along with monetary assistance to set up a scientific retreat at Arcueil. The Society of Arcueil was equipped with the most modern laboratories, and staffed by promising young students. Some of Europe's brightest minds became members, including the naturalist Alexander von Humboldt, the mathematician Pierre-Simon de Laplace, the chemist Jean-Baptiste Biot, and Étienne-Louis Malus, whose studies on light were cut off when he died young, weakened by disease contracted in Egypt.

The Society paid no attention to the haul from Egypt, confining itself to hard science, but Berthollet surrounded himself with Egyptian icons. He had a sacred ibis on his newly minted coat of arms, and designed an office fit for a pharaonic priest. The furniture was carved with columns, Sphinxes, and pyramids. The drapery was woven with lotuses, stars, and flowers, after the paintings in the tombs. His desk chair was an exact re-creation of the chairs in the ancient paintings. Panoramic views of the temples of Dendara and Thebes decorated the walls. His desk was sculpted on four sides to mimic an Egyptian temple and built with shelves to receive all the volumes of the *savants'* great book on Egypt.

Despite the prolific references to pagan spirituality around him, Berthollet remained a materialist, a deeply scientific man, rational and utterly without religion. Napoleon never really liked this aspect of his personality, and found him a bit chilly, although the emperor loved the chemist for his abilities. For companionship he preferred the passionate geometer Monge, but he respected Berthollet, a man who could help him make gunpowder and steel, maybe even sugar when it was scarce. The emperor not only made Berthollet a count, he made him a senator, a burden the apolitical chemist bore without complaint.

Berthollet would have preferred to leave politics, and all the messy aspects of life itself, including love and death, at the gates of his Arcadian compound. In spring 1810, however, his only son, Amédée, also a chemist, committed suicide, at age thirty, unexpectedly and for reasons unknown, but in a particularly scientific way. He plugged up all the holes in a room, lit a coal stove, and thus slowly asphyxiated himself, making notes in a notebook on all the symptoms of his approaching death.

Berthollet was in Paris when he received the news. He could intuit complicated chemical processes, but he was no man of words. He could not even find a way to tell his wife that their only son had died and asked a friend to do it for him.

A few months later, he wrote a one-line note to the English scientist Charles Blagden in London. "M. Chenevix has duly informed you of the disastrous loss of my son; thus it is necessary to resolve oneself to remain alone."

Berthollet survived the upheaval of Napoleon's fall professionally unscathed, and he would give the eulogy for his friend Monge, whose later years were not so easy.

Napoleon made Monge the Count of Péluse—after Pelusium, ancient site in the Sinai desert where the Persians surrendered to Alexander the Great. Along with the title came a fine château. For a time, Monge, his wife, and daughters enjoyed posh lives formerly reserved for hereditary nobility. The ardent revolutionary had scorned all trappings of nobility when he was a peddler's son, but in age and wisdom he graciously accepted these honors as his due.

In winter of 1813, the geometer's fortunes, which could be charted in a line parallel to Napoleon's, arrived at a steep angle, heading straight down. He was sixty-six years old when he read Napoleon's Bulletin 29, telling of the French army's disastrous retreat from Russia. Monge had a stroke of apoplexy. Recovering in a haze of smelling salts and brandy, he says: "A little while ago I did not know something I know now; I know how I shall die."

He was right in predicting that a stroke would be the thing to kill him. On that summer afternoon, though, he could only see the blurry outlines of his fate, not the details. Meanwhile, the emperor's grandiose plans waxed as his power waned. Napoleon wanted to scientifically assay all of North and South America, a project to begin after he had pacified all of Europe. "I need a companion to put me abreast of the present state of the sciences," the emperor said to Monge. "Then you and I will traverse the whole continent, from Canada to Cape Horn; and in this immense journey we shall study all those prodigious phenomena of terrestrial physics on which the scientific world has not pronounced its verdict."

"Sire, your collaborator is found," Monge replied. "I will go with you!"

Napoleon was not convinced. "You are too old, Monge. I need a younger man." Monge, ever game, went to work to find a younger man. While he was thus engaged, the British sent the emperor off to exile in the South Atlantic.

The restored monarchy stripped the Count of Péluse of both his title and his country estate. In 1816, the new regime ordered the nation's scientists to expel the great geometer from the national Academy of Science. Monge in his last years knew the same kind of poverty he'd experienced as a boy. He moved from slum to slum, living hand to mouth off the crumbs passed to him by scientific colleagues less reviled by the royals.

When he finally died, in 1818, of a stroke, his death passed without public homage, although Berthollet gave a eulogy in his quiet, unemotional way. As final punishment, the king forbade the students at the École Polytechnique to attend Monge's funeral. The following day, however, the students marched en masse to the Père-Lachaise cemetery, where they laid a wreath at the grave of one of their school's chief founders.

Like Berthollet, Geoffroy Saint-Hilaire and some of the other

French scientific heroes of the Egyptian Institute, Monge is immortalized with a street bearing his name in Paris. The rue Monge is a long, broad avenue that connects the 7th and 5th arrondissements, linking the intellectual quarter of the Museum of Natural History and the Sorbonne with the fashionable and commercial sector of the Left Bank.

Some of the scientists returned to France with lasting health problems that shortened their lives. Nicolas Conté returned to France robust, but his beloved collaborator and younger brother died in 1802, and his wife died a few years later. Those deaths broke him, and he himself died of an aneurysm at age fifty in Paris in 1805. His biographer, Jomard, attributed the inventor's early death both to the hard years in Egypt and his broken heart. Conté kept his ideas in his head and rarely wrote them down, so there is little evidence of what he would have produced had he survived. He left behind a single notebook, which has since been lost. Some of his inventions can still be seen at the school he founded, the Conservatoire des Arts et Métiers in Paris.

Fourier's career rose and fell with Napoleon's. As prefect of Grenoble, Fourier drained marshlands and eradicated malaria in his district. From the post, he also mentored a young man named Jean-François Champollion, who would go on to decipher the Egyptian hieroglyphic writing.

When the royals were restored, they promptly fired Fourier, and he spent several years in Paris selling his belongings in order to feed himself. Scientists at the Academy, ordered in 1816 not to aid the emperor's friend, defied the king and made Fourier their permanent secretary. From that chair, he regaled younger members with memories of Egypt, until some of them dubbed him "an insufferable bore."

On May 16, 1830, Fourier died, after falling down the stairs in his apartment. Twenty years after his death, Fourier's hometown of Auxerre erected a bronze statue to the brilliant, orphaned tailor's son. The Nazis melted it down during World War II.

The aristocratic doctor Desgenettes didn't always agree with Napoleon, but, like Fourier, his fortunes were tied to those of the little general. When he got back to Egypt, Napoleon awarded him the rank of baronet, for which Desgenettes, who had already discarded one title, had little use. Still, he loyally served Napoleon's armies to the bitter end, seeing action from Austria to Russia.

Alone among "the Egyptians," Desgenettes did not surround himself with Egyptian souvenirs and relics, nor did he bore his listeners with repeated memories of Egypt. Rather, what he did carry from those years was seared on his soul, and emerged only once—during an episode of great hardship during the last of Napoleon's wars in Europe.

In 1812, Desgenettes was with a garrison of troops in Torgau, in what is now eastern Germany. The weather was cold, raw, and wet with snow, and the soldiers were suffering from the usual maladies of an army in retreat: hunger, frostbite, lice. There was something else afoot, though, some sort of contagion that killed. Without medicine, or even a diagnosis, the number of the dying grew daily. The medical officers were overworked, and a large number of them, too, had already died.

Desgenettes was exhausted, feverish himself, and apparently sensed an awful, overwhelming familiarity in the stench and starvation. He called in all the civil doctors, and tensely demanded that they tell him immediately if any of their patients have buboes, or any of a list of other possible symptoms of *la peste*. The civil doctors raised their eyebrows at each other but agreed to look for the telltale signs. What the soldiers were dying of, most likely, was simple flu. Bubonic plague was the least of their problems. It was a ghost that haunted only the chief doctor.

By December, the Cossacks overran Torgau and hauled Desgenettes off to a cell. He appealed, in writing, directly to Czar Alexander, reminding the Russian leader that he was a medic who has performed numerous acts of mercy on behalf of sick Russians. The letter, remarkably, worked, and a contingent of Cossacks escorted Desgenettes back to French-controlled territory. He personally witnessed Napoleon's defeat at Waterloo, and watched the dethroned emperor leave for his final exile on St. Helena.

Napoleon always considered Desgenettes a bit too independent and cool, but the Bourbons distrusted the emperor's doctor even more. When they regained power, they kicked him out of the army. The doctor spent the rest of his life teaching medicine. He never lacked for confidence, and he eventually published his own medical memoir of the Egyptian campaign. Late in life, he was equally known for his wit and his healing skills.

When the scholars returned to France, Vivant Denon's slim, two-volume book, *Travels in Upper and Lower Egypt*, was already in press and about to become a smash hit across Europe. Denon wrote his book for popular consumption. It was smaller and more easily accessible than what the scholars planned or could do as a group, as Denon was the first to admit. It went into forty printings, was translated into Italian, German, and English, and became the nineteenth century's first best-selling travelogue. Denon's success inspired some other *savants* to try to personally publish their own works. None succeeded as he did.

Denon always deferred to the Commission and its superior analytical skills, though. When the scholars returned home safely, Denon took much of his own speculation out of the book, insisting that the scholars' interpretations would be more accurate. "I removed from my journal all that I had ventured in the way of research material; I put on again my uniform of soldier-scout . . . so as to guide those who were to pick up my footsteps and, were it only through my errors, to be of service to editors of the important work." The scholars didn't all respond with grace to the old diplomat's modesty. Devilliers, writing to Jollois in early 1802, peevishly belittled Denon's efforts. "Denon's zodiac was just as we saw it in Denderah, that is, it seems very small and from my point of view, very incorrect," he wrote. One of his École Polytechnique professors had fanned Devilliers's pride, telling the young men that their work was "more telling than Denon's" and that a separate collection of their drawings would be "magnificent."

Napoleon appointed Denon to be the first director of the museum of the Louvre. Denon became, in effect, the emperor's personal art consultant, cataloguing and arranging in the new national museum works of art snatched from wars all over Europe. The emperor himself had no feeling for fine art, a fact that embarrassed him. When shown a great masterpiece at a salon, he invariably replied, impassively, *"De qui est-ce?"* He said the great pyramids in Egypt were the most impressive works of art he'd ever seen.

Denon was an aesthete and could cover for Napoleon. He also understood, without needing to be told, the emperor's need for visual propaganda. Well into his sixties, Denon was still fearless on the battlefield, always willing to go to the latest European front and sketch, so that the artist to whom the job of painting the scene was finally assigned would know how it really looked at Austerlitz. Denon would describe exactly what color eyes the heroes on the field had. Denon knew which artists to commission for these jobs, just as he knew which of the looted paintings from Italy should hang in the national museum, and which should remain in storage.

In Paris, he existed in perfumed, elegant comfort, surrounded by witty friends and sensual delights. Courtier, diplomat, artist, writer, lover, he lived a full life of the mind and the body. In the years after his return from Egypt, Denon not only published his wildly popular travelogue, he tried his hand at literature, and succeeded. In just twenty-four hours he produced a first-rate short story after taking a bet that he could write a realistic love story without using obscenity. The result, titled "Le Point de lendemain," received rave reviews. The novelist Honoré de Balzac called it a primer for married men and "an excellent picture of the customs of the last century."

In 1815, the royals suggested he resign his post at the Louvre. Denon took his books and his puckish self-portrait, leaving behind for the museum's collection his mummy's foot and papyri. A dilettante to the core, he soon found something else to do with his mind and talents. For the next thirteen years Denon worked on an ambitious book on modern and ancient art. In it he planned to identify the single essential thread of human culture, a thread that linked the ancient Egyptians to the Greeks, to the Romans, to the painters of Byzantium, Venice, Bruges, to Michelangelo and Titian, and finally to the French painters of the nineteenth century.

In 1827, he died, leaving the work unfinished.

In 1830, three years after Denon's demise, Geoffroy Saint-Hilaire and Cuvier, by then two grizzled old naturalists, had a famous shouting match in front of the French Academy of Science's illustrious members. In fact, they argued in public every day for weeks. What had begun as a highly esoteric debate over whether form or function determined the phenomena of life had deteriorated into a brutal verbal brawl. The ferocity of the debaters was so disturbing and so shocking that the Academy's directors finally stopped it, closing the topic to future discussion.

Geoffroy Saint-Hilaire, who had truly lived for Cuvier's respect and begged for his letters while in Egypt, was immune by this point to Cuvier's disdain. He had his own supporters and his own fame. He was more than a scientist, he was a philosopher. Common people in France knew him as the man who personally led the first giraffe ever seen in Europe—a gift from the new Egyptian leader Muhammad Ali—on the road from Marseille to Paris in 1826. What's more, great artists and writers adopted and revered him as their patron philosopher.

Returning to France with his theory of the unity of life coursing through his head, he kept quiet about it at first, realizing that his colleagues would not approve. Instead, he worked out a more acceptable theory of animal organization, trying to demonstrate that all verte-brates—fish, reptiles, birds, mammals—shared a general plan of orga-nization, provable by their analogous parts. He first expressed the idea in 1806, when he wrote an entry on the Egyptian puffer fish for *The Description of Egypt*. In it, he proposed that a large, yet unnamed, bone in the fish was homologous to the human scapula. This line of reasoning eventually grew into a theory he called "the unity of composition," which got him admitted to the prestigious Institute of France (later renamed the Royal Academy of Science), and won him a second pro-fessorship.

Soon after, he expanded his idea to propose a more speculative theory about a single archetypal animal as the basis for all life. In 1817, Geoffroy published his grand theory, under the weighty title *Anatomical Philosophy: Of the Respiratory Organs with Respect to the Determination and the Identity of their Osseous Pieces*. He opened it with a long polemical defense of his right to be controversial, and then announced the epochal sig-nificance of his idea. This radical, unprovable speculation provoked the final breach with Cuvier.

Geoffroy Saint-Hilaire and Cuvier's bitter public debate was a bit personal in origin (the "dear friend" had never once reached out to his young protégé in Egypt), but it was also a debate about attitudes toward science, speculation versus certainty, Enlightenment versus Romantic. Cuvier was skeptical and formal, devoted to facts, while Geoffroy Saint-Hilaire was unpredictable, experimental. Simplified, theirs was also an epochal clash between two ways of looking at the world.

Geoffroy's search for an all-encompassing life theory made him popular with European writers and thinkers who were rejecting the pedantic, classifying style of the eighteenth century and seeking a more transcendent worldview. One of Geoffroy's many literary fans was the poet and naturalist Johann Wolfgang von Goethe, who, upon hearing of the public clash between Cuvier and Geoffroy, was beside himself. "The volcano has come to an eruption; everything is in flames, and we no longer have a transaction behind closed doors," he swooned to a friend.

Geoffroy Saint-Hilaire believed that with his theory of the unity of life he had fulfilled Napoleon's desire, expressed in the Cairo garden to Monge before he fled Egypt, to discover the universal laws governing the action of the smallest particles of matter. The zoologist dedicated his book on the unity of life to Napoleon.

By the time the novelist George Sand met him at the Jardin des Plantes in 1836, Geoffroy was thoroughly eccentric. "The old Geoffroy Saint-Hilaire is for his part a rather curious beast, as ugly as the orangutan, as talkative as a magpie, but for all that full of genius," she wrote. Sand, like other literary Romantics, was a great admirer of the old zoologist. "I have found in [his] new perception of Creation that which is most worthy of faith, that which is most satisfying to the human spirit . . . and to man's inexhaustible thirst for order and harmony: the universal unbroken chain, the balanced and harmonious joining by innumerable links . . . of the stone to the plant, of the insect to the bird, of the brute to the man, of man to the all, and of the all to God."

Balzac, another literary admirer, used Geoffroy Saint-Hilaire's unity of the species to create a "social species" in his *La Comédie humaine*. He credited Geoffroy Saint-Hilaire in his foreword, stating that his theory of the unity of composition was more than a "scientific innovation," it was a subject on which the greatest mystical writers of the age and the "finest geniuses of natural history" had meditated.

Like Savigny, Geoffroy Saint-Hilaire went blind, although late in life. He died in June 1844 and was eulogized as a pantheist, mystic, and humanist. Numerous literary figures attended his funeral. Two decades later, Darwin cited him as one of the men whose thinking anticipated the theory of evolution.

The longest-lived of the Egyptians were of course, the students.

For them, too, nothing that happened later in France could ever compare to those three years. Not love, work, or death. They built bridges and roads, married, and grew old, passing their lives in the bright, burning memory of a distant place they would never see again.

Like soldiers who have experienced the most intense moments of their lives in battle, the young men grew into middle years and then old age never fully at home in the present, always most animated when talking about the past.

Life at home was humdrum in comparison.

Prosper Jollois married a woman named Amélie Soufflot, a descendant of the famous architect of the Panthéon, but the couple had no children. Jollois had a lifelong passion for history: he did research on France's past, including the history of Joan of Arc, the siege of Orléans by the English, and on Roman and Gallo-Roman antiques in Paris. He died June 24, 1842, after having agreed with his old friend, Édouard, that they would be buried side by side.

During the last years of his life, Édouard Devilliers, who never married, also had a passion. The old bachelor remained devoted to unlocking the secrets of ancient Egyptian astronomy, though he never contributed anything further to the canon of Egyptology. He died in 1855.

On an early spring morning in April 1855, Jomard, the last of the Egyptians, eulogized Devilliers and Jollois together at their adjacent graves in Paris with a windy, romantic speech. "Two young men met one day during the memorable expedition to Egypt; they became bound by a strong affection founded on esteem; they became inseparable; their names were only one name; only death could separate them. *Adieu*, Édouard Devilliers; *adieu*, Jollois. You both, our companions on the trip, have preceded us by a few days to the shared rendezvous; you have served the homeland, your lives were full, you have left names that will never perish. Each of you could say to yourselves on leaving the earth: '*Non omnis moriar, multaque pars mei vitabit Libitinam.*' "

The Latin is the last line of a verse by Horace, which most of the educated listeners of the time would have recognized instantly. The full verse, in translation, is:

> *I have finished a monument more lasting than*
> *bronze and loftier than the Pyramids,*
> *One that neither wasting rain, nor furious north*
> *wind destroy, nor the years erode.*
> *I shall not altogether die, but a mighty part of me*
> *shall escape death.*
> *On and on shall I live and my reputation grow,*
> *ever fresh with the glory of time.*

FROM EGYPTOMANIA TO EGYPTOLOGY

It would be hardly respectable, in one's return from Egypt, to present oneself in Europe without a mummy in one hand and a crocodile in the other.

—The monk Father Germab, to the
Egyptian ruler Muhammad Ali in 1833

The French scholars and artists returned from Egypt believing they were bringing home a body of knowledge that would edify their colleagues about animals, plants, minerals, and medicine from the distant land, and increase Europe's general understanding of the culture of ancient and modern Egypt. None of them would have predicted that the chief consequence of their efforts would be on European fashion, art, and architecture. Nor could they have known that a secondary result of their work would be to inspire a century of wholesale cultural plunder, an insatiable European demand for things Egyptian that they, with a great book that owed its existence to an invasion, had helped create.

The French had already ushered in the phenomenon known as Egyptomania before the campaign with their use of Egyptian iconography in revolutionary art. After the invasion, they went even further. Of course, they struck medals to the campaign's war heroes decorated with Egyptian motifs. They erected massive columns (such as the Fontaine de la Victoire in the Place du Châtelet) in public squares with palm capitals. Napoleon himself chose as his personal heraldic emblem, not the traditional fleur-de-lys, but the bee, which classical authors believed was the hieroglyphic symbol for "ruler." Pyramids and other Egyptian elements fill Paris's Père-Lachaise cemetery designed under Napoleon. Several of the *savants* are buried there.

Denon directed the great French porcelain maker Sèvres in 1804 to create an elaborate dinner set that became known as the Egyptian Service. Decorated with pyramids and lotuses and replete with Sphinx and obelisk sugar bowls and creamers, these fantastic black-and-gold plates, teacups, and saucers replicated, improved upon, and miniaturized the images from *The Description of Egypt*. It was meant for Joséphine but ended up with the Duke of Wellington, a piece of booty that crossed the Channel thanks to Napoleon's fall.

Even Napoleon's final defeat was linked to his Egyptian dream. The French had planned to commemorate the Battle of Waterloo—if a success—with a pyramid of bricks covered in white stones.

Other European craftsmen and artists followed the trend. Before the war in Egypt was even over, English players staged an Egyptian operatic festival, with lavish Egyptian-style sets and costumes. Italian jewelers created scarab brooches. Furniture designers from many countries made desks, chairs, and tables playing with Egyptian shapes (obelisk) and motifs (winged stars, lotus, Sphinx, crocodile, Isis and Hathor heads). The longest-lived piece of furniture with an Egyptian element may be the metronome, designed during this period and placed inside a truncated obelisk. Winged globes, pyramids, palm- and lotus-headed columns—Egyptianisms, as the design world calls them—soon appeared everywhere, in libraries, on storefronts, on chimneypieces.

In England, the poet Robert Southey complained about the lurid Egyptomania inspired by soldiers returning from service in Egypt. "The ladies wear crocodile ornaments, and you sit upon a sphinx in a room hung round with mummies, and the long black, lean-armed long-nosed hieroglyphical men are enough to make the children afraid to go to bed. The very shop-boards must be metamorphosed into the mode, and painted in Egyptian letters which, as the Egyptians had no letters, you will doubtless conceive must be curious."

It is probably not a stretch to say that one of the strongest elements running through the design of the early and middle nineteenth century were rooted in images of Egypt. By the middle of the century, no world's fair was complete without an Egyptian room, as the historian Donald Reid has pointed out. Verdi's *Aïda*—about the travails of an Ethiopian slave in pharaonic times—was one of the highbrow excesses of the trend.

The fad also extended to a desire among collectors to own a real piece of Egypt. To feed this craving, a thriving tomb-raiding business developed, in which thousands of objects, from the tiny to the colossal, made their way out of Egypt and into the homes, mansions, and museums of Europe. Very soon, this plunder made the *savants'* collection of sarcophagus, Rosetta Stone, and colossal fist, look picayune.

The rape of the Nile, as it's been called by various writers, was probably at least partly provoked by the reports of the *savants* and the tantalizing and beautiful images in their book, although it might well have happened anyway as a by-product of European colonial expansion. The *savants* themselves did not haul much booty back to France, though not for want of trying. The seven-ton torso of Ramses II—now in the British Museum—has a hole bored in its chest by French scholars in a futile effort to figure out how to move it.

The wholesale plunder of ancient Egypt began in earnest a few years after the *savants* left, under the stewardship of a dictator, Muhammad Ali, who took control of the country after the French, English, and Turks left. Muhammad Ali was an orphan from Macedonia who first entered Egypt with the Ottoman force sent to battle the French. He built a power base of Mamelukes and villagers and ruled Egypt with an iron fist from 1805 to 1849. While the country remained nominally within the Ottoman Empire, Muhammad Ali set it on a separate course from Turkey for the first time, and made Egypt a nation of its own. This nascent independence grew until 1882, when the British invaded Egypt, opening an era of English control that nonetheless failed to eradicate the dominant French cultural influence planted there by Napoleon.

Muhammad Ali centralized the national administration and sent doctors, soldiers, engineers, and educators to Europe—often to France—for training. He built a new agricultural base by extending the cultivation of cotton and expanding irrigation. He extended Egyptian power into the Sudan, Syria, and Arabia, though he would be forced by the Europeans to give up Syria and Arabia.

Throughout the first half of the nineteenth century, the new pasha and his familial heirs worked to bring European technology and capital into Egypt. They opened the trade routes and gave Europe access to raw materials. They traded raw cotton, but that wasn't all. They also benignly looked away as Europeans took home tons and tons of ancient artifacts, made easier to transport with each advance of the Industrial Revolution.

The first post-Napoleonic French and British consuls in Egypt, Bernardino Drovetti and Henry Salt, were enthusiastic participants in the new antiquities trade, which grew in the context of European national rivalry. By the 1820s, the Nile valley was buzzing with diggers operating in the major sites, from Giza to Philae, under excavation. The race to find, claim, and bring objects back to European capitals was on.

The title of "greatest plunderer of them all" initially went to an Italian weight-lifter and actor named Giovanni Battista Belzoni. Belzoni, a six-foot-six Italian giant, began his professional life as the base in an inverted human-pyramid act, performing in venues in London, where he astonished audiences by holding aloft more than twenty people at a time. While traveling through Europe, Belzoni befriended an agent of Muhammad Ali and somehow persuaded the Egyptian that he could build an extremely efficient waterwheel to revolutionize agriculture along the Nile. Soon Belzoni was deeply involved in the Egyptian antiquities trade. He started by using his brute strength and moderate mechanical skills to organize the removal of a colossal head from the sands of Thebes, haul it upriver and across the sea to Europe and its final resting place in the Egyptian wing of the British Museum. Thus began a long and profitable career for Belzoni of brushing away sand and moving antiquities large and small from sites all along the Nile.

Between them, Salt, Belzoni, and Drovetti competed with each other to tear away any loose objects from sites like Thebes and ship them to Europe—all under the bribed eyes of the pasha and his agents. By the time they died (Drovetti in an insane asylum, Belzoni of dysentery in West Africa, Salt of an intestinal infection) they were replicated in the thousands by dealers, collectors, tourists, and amateurs who began "a rape of massive proportions" in the mid-nineteenth century that continued for a hundred years or more. "The museums of Europe were now so eager to obtain Egyptian antiquities that they were quite prepared to ship entire rooms, friezes, or tombs," Brian Fagan, author of *Rape of the Nile*, has written.

Thanks to these efforts, the average citizen in New York, London, or Paris is familiar with such objects as the Dendara zodiac (dynamited from its place in the temple in 1821 and moved to the Louvre), Cleopatra's Needle (given to New York City as a gift from the French people—now standing in Central Park), the Luxor temple obelisk in the Place de la Concorde, and room after room of mummy and treasure in the British Museum, New York's Metropolitan Museum of Art, and the Louvre.

The sites were desecrated by all this plundering, although scholarly excavations of a sort were beginning. The Egyptians themselves were not yet concerned with historical preservation. Villagers aided in the foreign plunder and the government regarded the sites as natural resources of a totally nonhistoric sort. At Esna, where sand covered colossal temple columns nearly to their lotus cornices, Muhammad Ali began a dig, not with any "spirit of antiquarian zeal," wrote the nineteenth-century English novelist and traveler Amelia Edwards, "but in order to provide a safe underground magazine for gunpowder." By 1835, a quarter of the temple of Dendara had been quarried away for saltpeter. Other temples were taken apart for building stone.

One of the first voices to call for a stop to the destruction was an American, George Robins Gliddon, a former diplomat in Egypt who became a writer and lecturer on ancient Egypt. Since the Americans had as yet no vital stake in the colonial game being played by Europeans in the region, they were comfortably able to stake out the moral high ground. In 1849 Gliddon published a volume titled *An Appeal to the Antiquaries of Europe on the Destruction of the Monuments of Egypt*. The appeal was ignored.

At mid-century, a French scholar, Auguste Mariette, stepped in where Belzoni, Salt, and Drovetti had left off. Mariette, who had studied the Egyptian civilization from Paris, was appointed by Muhammad Ali to be the first director of the first Cairo museum. His finds, however, found their way mainly to the Louvre. He started out excavating tombs near Cairo and produced huge caches of gold and bronze. Men under his direction were eventually digging at thirty-seven sites up and down the Nile valley (he was authorized to requisition able-bodied village men). The excavations were notoriously careless and utterly unscientific. "Dynamite was among his techniques," notes the historian Fagan. "Careful recording and observation mattered nothing, only objects." In any case, Mariette was seen as a scholar back home, and that is how he remains in most history books. The catalogue for the huge Egyptian exhibit at the 1867 Paris Exposition said of Mariette that he "rescued ancient Egypt for Europe."

In the twentieth century, English archaeologist Howard Carter excavated at the Valley of the Kings, where he descended into the most famous, crowd-pleasing, Egyptian find of all, the gold-filled tomb of King Tut. The English gave this treasure to the Cairo Museum. This find opened the modern episode of the 5,000-year history of ancient Egypt and her relics, in which the inhabitants of Egypt are now seen as the primary caretakers of their cultural heritage.

The treasures of ancient Egypt remain scattered throughout the museums of the world, although there is an effort under way by Egyptian cultural authorities to reclaim some of it. In 2003, the Egyptians demanded the British return the Rosetta Stone to Egypt, "if they want to restore their reputation." The British, as of this writing, have not moved the stone and show no inclination to do so.

Much has been written about the implications of the plunder of Egypt. Western scholars now tend to express a certain amount of remorse over their heedless predecessors, while writers like Edward Said and other critics of colonialism see archaeologists over the years as witting or unwitting henchmen for the imperialist, European powers and their "Orientalist project." The politically incorrect view, of course, is the one emblazoned on the 1867 Paris Expo catalogue, celebrating the Europeans who had "rescued" the ancient culture from sand, garbage, and time.

The French scholars deserve to be credited with some positives. After they left, Egypt had the wheelbarrow, the printing press, and a new emphasis on foreign languages and professional education, the latter much promoted by the Eurocentric new pasha. The French also left in Egypt a lasting love of the French language, such that well into the 1980s one could find Egyptian elites who believed that fluency in French separated the boors from the sophisticates. Until quite recently, street signs in Cairo were written in both Arabic and French.

Scholars from different political and geographical points of view have debated the following questions over the years: Did the scholars play any role in increasing or ameliorating the mistrust between Islam and the West that so plagues the modern world? Did they help bring

Egypt out of the Middle Ages (or have no effect)? Was their book a bridge between the two cultures, or merely a Western stamp of cultural lordship over an Eastern territory?

The gigantic linen pages of *The Description of Egypt* attest to the fact that the scholars did their best, within the limitations of knowledge and cultural sensitivity in the early nineteenth century, to give Egypt its due. After three years of hunger, hardship, uncertainty, and disease, they returned to France imbued with a powerful respect—and in some, a real affection—for the land in which they had been stranded. Their great book, coinciding with an Industrial Revolution that facilitated the transportation of colossal artifacts to Europe, has been unfairly blamed for the cultural plunder that followed. Napoleon's scientists returned to Europe with their sight, their health, in some cases their thought processes, permanently impaired, and yet so glorified the country of their temporary exile that future generations were moved to imitate, to touch, and to see it for themselves.

NOTES △△△△△△△△△△△△

THE GENERAL

1 *There was nothing more surprising* Bernoyer, 19.

2 *"Thousands of men leaving"* Denon and Kendall, Vol. 1, 2. Scene at Toulon relies on various accounts, including Reybaud et al., Vol. 3, 59–62; and Laissus, 62. Hymn to liberty: Bainville, 93. Napoleon attending beheading ceremony with scientists: Bainville, 94. Shipboard salons: Laissus, 65. Numbers and types of members of the *corps de savants*: Goby, 290–301.

5 *"Everyone is talking about* Geoffroy Saint-Hilaire (Étienne), 30.

5 *"When they went to eat* Bernoyer, 19–20.

7 *always keen to "get a whiff of gunpowder"* E. T. Bell, *Men of Mathematics*, 194.

7 *"Berthollet has a rather ordinary exterior,"* Sadoun-Goupil, 37. On the wearing of togas: Byrd, 56. Much of the discussion about the state of the Ottoman Empire and its relations with France here and throughout relies on Lord Kinross, *The Ottoman Centuries.* Napoleon's saying he would happily trade all of Italy itself: Herold, 13. Explanation of colonial competition between European powers in 1800 owes much to Maya Jasanoff and Christopher Herold.

12 *"that most savage and unprincipled nation,"* Jasanoff, 118.

12 *"a conquest taken from the English."* Jasanoff, 132.

13 *"They stripped them naked,"* Volney, 210.

15 *"Soldiers!" it began.* Official papers, No. 1.

17 *"The merchants remain at Cairo* Volney, 214.

THE GEOMETER AND THE CHEMIST

19 *"Monge loved me* Quoted in Bell, *Men of Mathematics,* 190. Descriptions of life in France in the 1790s are based on Lyons.

23 "The scientists who enlisted" Details of Bertholet's work assembling expedition. Laissus, 35–36.

23 *Historians of the expedition believed* The idea that 50,000 men could be mustered for an expedition into the unknown seems unbelievable in the era of instant global communications, but no historical accounts refute it. Some of the scholars knew where they were headed, others had hunches, but the soldiers thought they were headed off to fight the British, and very few of them knew their actual destination was Egypt until well into the sea voyage.

24 *"They speak of Egypt."* Laissus, 53. Denon's life story from Nowinski. Details of the disembarkation at Alexandria are from Bernoyer, Laissus, Herold, Denon, and the journals of various *savants.* The horrific desert march has been described in many books and journals, including Herold, Laissus, Bourrienne, Bernoyer, and the Dible and Richardson biographies of Larrey.

24 *"My imagination went back to the past,"* Denon and Kendall, Vol. 1, 15–16.

28 *"I walked among the soldiers* Bernoyer, 51.

28 *"Marching on foot* Richardson, 24. Monge's letter comparing Napoleon to Jason: Laissus, 42. Monge's life story is entertainingly told by Bell.

31 *"Descriptive geometry is a language* Gillispie, 524.

32 *"I am not very young.* Laissus, 40–42.

32 *"Old fool," she fumed,* Laissus, 40–42. Descriptions of Mameluke dress, social customs, and military tools and tactics come from a variety of sources, including Volney, Bruce, Ayalon (Outsiders in the Land) and Herold.

33 *"Strangers to each other,"* Volney Vol. 1, 173. Descriptions of the chief beys are from Herold and Jabarti.

37 *L'eau du Nile* Laissus, 86. Descriptions of the Battle of the Pyramids: Laissus and Herold.

39 *"The General enters Cairo* Bourrienne, 122.

THE INVENTOR

41 *"It was not variegated* Denon and Kendall, Vol. 1, 14.

41 *"We were looking for the city* Laissus, 78. Devilliers's crisis of confidence: Devilliers, 39.

42 *"The first scene that presented itself* Denon, Vol. 1, 20.

43 *"In its dogs, he could not recognize* Denon and Kendall, Vol. 1, 21.

44 Biscuit anecdote Devilliers, 42. Comparison between military supply tactics in Italy and Egypt: Herold. Discussion of ophthalmia, fevers, insects, and dysentery appears in all the journals.

44 *"They were already busying themselves* Jollois, 43. Examination of the pillar, discovery of the sarcophagus and ancient sheet music: Laissus, 80–81.

47 *"Only the Arabs' humanity* Bernoyer, 48. An overview on how the savants felt about Islam, see *La Description, Etat Moderne.*

47 *"this bizarre spectacle* Laissus 92. For Conté's life story, see Jomard.

51 *"It is an earthly paradise!"* Geoffroy (Étienne), 79. Napoleon's intentions regarding Abukir and moving the ships: Bourrienne, 125. Eyewitness accounts of the Abukir Battle and its aftermath come from many scholars' journals but are especially vivid in those of Devilliers and Jollois.

53 *"The night was dark.* Jollois, 51–52.

53 *"Apparently the French have no more fleet!"* Devilliers, 59.

54 *"Our beasts of burden shuddered* Laissus, 95.

54 *"The odor that reigns* Jollois, 54.

54 *"From then on we realized* Malus, 89.

55 Berthollet called Conté *"the column of the expedition* Jomard, epigraph.

56 *By means of this device . . . 428 feet* Jomard, 31.

56 *"He had to render each scene* Jomard, 35.

57 *"What are we going to do now?* Jomard, 26–27.

THE INSTITUTE

59 "O Egyptians!" Bonaparte's Proclamations, 5–6. Scene of the first meeting of the Institute is amalgamated from accounts in Laissus, Herold, Geoffroy (Étienne), and author's visit to the restored building in Cairo. Descriptions of Cairo in 1800 owe much to Anderson and Fawzy, Hourani, Raymond, and Rodenbeck.

64 "This [order] was in spite Jabarti, 66.

65 "Inhabitants of Egypt!" Proclamations, 7.

65 "I arrived at 11 o'clock Laissus, 128–29.

66 "to prove to the inhabitants Jomard, 58. Tricolored rice: Herold, 153.

67 "Things are even better in Cairo Geoffroy (Étienne), 69.

67 "It is very funny to see . . . Frenchmen Geoffroy (Étienne), 62. Cairo civic organization: Hourani, 235–47.

68 "the metropolis of the universe . . . stars of knowledge." Anderson and Fawzy, "Medieval and Modern Egypt," Hilary Weir, 90.

69 wa'ad al bisul—"promises of arrival." Said, 342.

73 "Here I once again find men Herold, 176.

73 "Besides the regular sessions of the Institute Herold, 171.

74 "I owe Providence Geoffroy (Étienne), 87. Rodenbeck details the Mamelukes' obsession with controlling written information.

76 "To calm the jealousy of the soldiers Jollois, 7. The Divan discussion relies on Laissus and Herold, 182ff. Anecdotes about Napoleon's dressing in Turkish costume: Bourrienne, Herold, and Las Cases.

79 "Will it be said that Las Cases, Vol. 1, 88.

79 "This was quackery Las Cases, Vol. 1, 88.

79 "Is the dead man your cousin?" Las Cases, Vol. 1, 87–88.

80 "What they promised did not come true. Hourani, 293.

80 "We were shown other experiments Herold, 172.

81 "The French installed a great library Jabarti, 109.

81 "The native population [at first] Herold, 173.

82 "Her figure was distinguished Bernoyer, 75–76.

83 "Everything is permitted to them," La Description, Vol. 18, 456–57.

84 "One fine day, some sheik or other Herold, 192.

84 *"Would you dare to abandon* Herold, 194.

85 *"The angel! The angel!"* Herold, 195.

86 *"They treated the books and Koranic volumes as trash"* Jabarti, 93.

THE ENGINEERS

92 *"We fight, we grumble* Goby, *Un compagnon de Bonaparte*, 7.

92 *In a closet-sized berth* Devilliers, 18–20.

92 *"One must hope that the Government* Jollois, 5.

93 *"Your sensitive and tender feelings* La Sabretache, 60.

93 *"All my companions suffer cruelly* La Sabretache, 60.

93 *"We are 110 in a room* Devilliers, 31. Student attitudes toward their military rank and the soldiers: Laissus, 57–58.

94 *"la maîtresse favorite du général"* Jollois, 7; also Geoffroy (Étienne), 27–28.

95 *"I will try hard* Devilliers, 33.

95 *"I swam across the Nile* Devilliers, 33.

96 *"The façades of the houses* Devilliers, 48.

96 *"I admit the repugnance I had* Jollois, 56.

97 *"Everyone is out at nine at night* Devilliers, 71–72.

97 *"I was supposed to get* Devilliers, 72–73.

98 *"I arrived halfway up* Jollois, 62. Devilliers and Dubois-Aymé's visit to pyramids: Devilliers, 76–78.

99 *"Neither . . . made any response!"* Devilliers, 79.

99 *"During his stay in Egypt* Sadoun-Goupil, 47.

99 *"When I find myself sad* La Sabretache, 67.

99 *"He doesn't care what happens* Las Cases, Vol. 1, 210. Details about the archaic way the engineers conducted surveys and their demographic work owe much to Charles Coulston Gillispie's historical introduction to *Monuments of Egypt.* Jomard on the doors: *La Description*, Vol. 18, 451–52.

102 *"The city is almost entirely* Anderson and Fawzy, 91.

103 *"In one public place, singers gather* La Description, Vol. 18, 439–40. Jomard on the a'lmeh: *La Description*, Vol. 18, 441–43. Jomard on the *fellahin*: *La Description*, Vol. 18, 438–39.

104 *The caution and pride attached to these documents* Godlewska, 16. The thumbnail history of how different cultures and religions viewed the Nile comes from Robert Wild, *Water in the Cultic Worship of Isis and Sarapis*, 95. History of ancient efforts to build a canal: Kinross, *Between Two Seas*, 5.

107 *"He will then cut the Isthmus* Kinross, *Between Two Seas*, 1. Anecdote about Napoleon protecting the sacred tree, and Monge's running discourse on the biblical Red Sea: Kinross, *Between Two Seas*, 3.

109 *"Everyone who had been to Suez* Jollois, 83–84.

109 *"Multiple excursions were organized,"* Jollois, 87.

110 *"Chief Engineer of All the Works Between the Two Seas."* Jollois, 87. Napoleon's dromedary regiment and his reliance on camel-stealing: Herold, 223.

110 *Their leveling telescopes . . . were marked in decimals* Gillispie and DeWachter, 12.

110 *The returning civilians . . . "went to find General Caffarelli* Jollois, 87.

111 *"The French, intimidated by the revolt* Jollois, 75–76. Devilliers outraged by the new calculations: Gillispie and DeWachter, 11.

THE DOCTORS

115 *A long train of loaded camels* Dubois-Aymé, in *The Description of Egypt*, Vol. 15, 181. Larrey's and Desgenettes's life stories are from two biographies: Richardson and Gazel.

118 *"He owed to the milieu* Gazel, 43.

122 *"Unworthy of being a French citizen* Herold, 209.

124 *"May God cause the upheaval* Kinross, *The Ottoman Centuries*, 424.

125 *"They insulted those . . . on horseback,* Bourrienne, 152.

125 *"If you only knew, my dearest Laville* Richardson, 53.

126 *"The tumult of the carnage* Malus, 136–37.

126 *"All that can be imagined* Bourrienne, 157.

126 *"The frenzied pillage* Malus, 139.

127 *"Immediately after the departure* Malus, 140

127 *"I had assiduously gone* Malus, 142.

128 *"Citizens, I beg you* Desgenettes, 55.

129 *"I thought that I should* Desgenettes, 51.

129 *"The Turks brought in above sixty heads,"* *The Siege of Acre*, Smith Letter, May 30, 1799.

132 *"I was alone, without strength* Malus, 144.

133 *". . . had me called* ,Desgenettes, 67.

134 *". . . it by vomiting* , Desgenettes, 67–8.

134 *"The utmost disorder* Smith letter, May 30, 1799. Napoleon's refusal to put a plague victim on his own horse: Bourrienne, 168.

135 *"After a few minutes of torture* Richardson, 35.

135 *"Our thirsty soldiers threw themselves* Richardson, 40.

135 *"It is said that the plague* Jollois, 78.

136 *"like the rest of the army* Geoffroy (Étienne), 119–20.

136 *"Dutertre would ask* Gillispie and DeWachter, 43.

136 *"Bonaparte, impatient at some discussion* Gazel, 71.

137 *Napoleon . . . "pale with rage."* Laissus, 253. Royer's and Napoleon's differing versions of the Jaffa euthanasia poisoning are also in Las Cases, Vol. 1, 195–202.

THE MATHEMATICIAN

139 *My friends, if we leave for France* This and subsequent quotes about and details of the scholars' departure from Egypt came from Laissus, 270–73, and Devilliers, 223–25. Monge preparing to blow up the ship: Bell, 196.

143 *"Scarcely was the flag* Denon and Kendall, Vol. 2, 282.

143 *"deported into the deserts of Arabia* Herold, 330.

144 *"I was favored with the realization* Denon and Kendall, Vol. 2, 277. Denon's gift of monkey to Joséphine: Laissus, 276.

145 *"He is incapable of organizing* Kléber letter, part of the "Napoleon in Egypt" exhibit at the Dahesh Museum, New York, 2006. Dolomieu anecdotes from Lacroix and Daressy, 122–23.

146 *"We reacted emotionally* Devilliers, 225.

147 *"I am bored to death* Devilliers, 261.

147 *"I dream of my family* Devilliers, 261–62.

147 *"The people are inflammable.* Hourani, 292. French atrocities: Devilliers, 234–38.

148 *"We often saw men* La Description, Vol. 15, 186–88.

149 *"She has lost her honor,"* Denon and Kendall, Vol. 1, 31.

149 *"They wanted gold,"* he wrote, Denon and Kendall, Vol. 1, 58.

149 *"Nothing disturbs them,"* Laissus, 92.

150 *"We saw buffaloes in the water* Laissus, 91. Captain Moiret's poignant account of his doomed love affair with an Egyptian woman: Moiret, throughout. Female Mamelukes dressing like men to attract men: Rodenbeck, 87.

150 *"They are above all addicted* Volney, Vol. 1, 173. Mameluke attitudes toward female virginity and women in general are explained by Afaf Lutfi al-Sayyid Marsot, *Women and Men in Late Eighteenth Century Egypt* (Austin: University of Texas Press, 1995), chapter 3. Nafisa al Bayda was profiled by Agnieszka Dobrowolska in *Al Ahram Weekly,* August 25–31, 2005.

153 *"They sent their children far away* Geoffroy (Étienne), 121–22. For Mameluke attachment to Cairo, see Ayalon, and Mameluke household organization and social hierachy is extensively covered by Marsot (cited above).

154 *beardless men were assumed to be slaves* Geoffroy (Étienne), 116.

155 *"I am living here very peacefully* Geoffroy (Étienne), 121–22. Bernoyer devotes pages to his extended adventures seeking the perfect slave woman for himself in Cairo: Bernoyer, 94–116.

156 *"That bugger has left us* Herold, 341.

156 *"If by next May* Herold, 343.

157 *"At first he did not want* Devilliers, 223–25.

157 *"Muslims, bow down before him!"* Herold, 344.

158 *"All turned to sparkling fire* Hourani, 298.

158 *"The poor* savants *of Cairo* Geoffroy (Étienne), 146.

158 *"a trunk nerve of mathematical physics."* Bell, 183.

159 *"Fourier's theorem is not only* Bell, 183.

160 *"Fourier's aim was to have* Geoffroy (Étienne), 216–17.

161 [Jollois] *"knew how to conciliate* Jollois, 9. Fourier's life story is in Bell, Arago, and Herivel.

163 *"my very dear and precious Conté"* Jomard, 32–33. British efforts to foment mutiny among the French: Moiret, 125.

163 *"Either we'll all die together* Herold, 351.

164 *"Apparently, it was atrocious,"* Devilliers, 256.

165 *"It is true that the troubles"* Moiret, 150.

165 *"Our situation has not gotten better* Geoffroy (Étienne), 148.

THE ARTIST

167 *This abandoned sanctuary* Denon and Kendall, Vol. 1, 215.

171 *"Desaix was wholly wrapped up* Herold, 228.

171 *"Sometimes he [Murad] pushed bravery* Herold, 231.

172 *"Their ferocity is equalled only* Herold, 245.

172 *"You will be loved by women* Nowinski, chapter 1.

173 *"The surgeons, destitute of drugs* Denon and Kendall, Vol. 1, 190.

173 *"The marks of this assassination* Denon and Kendall, Vol. 2, 80.

174 *"My friend, is this not an error* Denon and Kendall, Vol. 1, 156.

174 *"The mind may paint* Denon and Kendall, Vol. 2, 71.

175 *He wrote that "a cut* Denon and Kendall, Vol. 2, 4.

175 *"The pride of raising colossuses* Denon and Kendall, Vol. 2, 63.

175 *"simplicity that produces the great."* Denon and Kendall, Vol. 2, 160.

179 *"Night rides always have* Laissus, 293.

181 *"We believed that you thought* Devilliers, 180.

181 *"One can easily conceive* Laissus, 296.

182 *"Reason succumbed* Laissus, 297.

183 *"gripped them like nothing* Ceram, 138.

183 *"The sound seemed to be coming* Devilliers, 210.

THE NATURALIST

185 *"After a sojourn of eight days* Denon and Kendall, Vol. 2, 124.

185 *"When I arrived at the cultivated lands,* La Description, Vol. 11, 396–397.

187 *"I'm not very happy,"* Laissus, 56.

188 *"One of the people there* Jollois, 176–77.

190 *"While the soldiers* Geoffroy (Étienne), 127.

190 *"One cannot master the transports* Geoffroy (Isidore), 111.

191 *"That which took the savants* Geoffroy (Isidore), 88.

191 *"The findings he made* Pallary, 30.

193 *"While common to all sorts* Pallary, 48–49.

193 *"He brought back from . . . Upper Egypt* Pallary, 26–27. Modern ophthalmologists' speculations about Savigny's troubles are from Gillispie, *Science and Polity in France,* 588.

THE ZOOLOGIST

195 *My dear Cuvier:* Geoffroy (Étienne), 129–30.

196 *"The whole world is attacked* Cahn, 42.

196 *"Am I destined,* Geoffroy (Étienne), 144–45.

197 *"It is far too much to boast* Geoffroy (Étienne), 137.

198 *"Further into the expedition,"* Appel, 74.

199 *"It is known that nature* Appel, 86.

200 *"remained foul for a long time* Devilliers, 280–82.

200 *"Many soldiers called us* Devilliers, 269–70. Geoffroy's unsuccessful attempt to get the *savants* sent home: Byrd, 251.

201 *"It is false that he tried* Geoffroy (Isidore), 88–90.

202 *"Osmanlis' [Turks'] arrival* Jollois, 136.

202 *"At first we thought* Devilliers, 290.

203 *"I see him again,"* Devilliers, 290–92.

204 *"I had brought a mattress* Devilliers, 292–93.

204 *"Some of the excited soldiers* Geoffroy (Étienne), 200–1. Devilliers lost a box of minerals and antiquities: Devilliers, 293.

205 *"A veritable famine"* Geoffroy (Isidore), 99.

205 *"We finally set sail* Desgenettes, 75.

206 *"His friends worried about him* Geoffroy (Isidore), 104.

THE STONE

209 *I have just been informed* Herold, 387.

210 *"He was very sad* Devilliers, 295–96.

212 *"Pointers would not range* Otter, 342.

212 *"They said, perhaps the going* Otter, 343.

213 *"I have just been informed* Herold, 387.

213 *"politely but coldly,"* Geoffroy (Isidore), 105–6.

213 *"You are taking from us* Reybaud et al., Vol. 3, 38–39.

215 *"While I vegetate* Geoffroy (Étienne), 202.

216 *"The words 'Jamais on a autant pillé le monde!'* Otter, 343.

216 *"You want it, Monsieur le général?* Herold, 387–88.

216 *"carried off the stone* Parkinson, 10.

THE BOOK

219 *By deploring the fate of so many* Geoffroy (Étienne), 204. I relied on Herold for the final count of returnees and dead.

222 *"The fall of the Empire* Geoffroy (Isidore), 307.

227 *"In Egypt," he told . . . his scribes* Herold, 3–4.

227 *"My departure from Egypt* Las Cases, Vol. 1, 206.

227 *"The desert . . . had a peculiar influence* Las Cases, Vol. 1, 62. Description of Berthollet's office at Arcueil: Sadoun-Goupil, 49. Berthollet to Blagden: Sadoun-Goupil, 71. Anecdotes and quotes re Monge: Bell, 183–205. Anecdotes about Fourier's later life: Arago and Herivel. Anecdotes re Desgenettes: Gazel.

233 *"I removed from my journal* Denon and Kendall, Vol. 2, 284.

233 *"Denon's zodiac . . . in Denderah* Jollois, 26–27. Denon anecdotes: Nowicki. Details on the Geoffroy-Cuvier debate and on Geoffroy's later years: Appel.

237 *"I have found in [his] new perception* Appel, 190.

238 *"Two young men met* Introduction to Devilliers, xxii–xxiii.

EPILOGUE

241 *It would be hardly respectable* Fagan, 11.

242 *"The ladies wear crocodile ornaments* Curl, 130. The adventures of early Nile plunderers Salt, Drovetti, and Belzoni: Fagan.

245 *"The museums of Europe* Fagan, 251.

246 *"spirit of antiquarian zeal,"* Fagan, 273.

BIBLIOGRAPHY

SOURCES IN ENGLISH

Adams, Andrew Leith. *Notes of a Naturalist in the Nile Valley and Malta, a Narrative of Exploration and Research in Connection with the Natural History, Geology, and Archaeology of the Lower Nile and Maltese Islands.* Edinburgh: Edmonston & Douglas, 1870.

Adams, Percy G. *Travellers and Travel Liars, 1660–1800.* Berkeley: University of California Press, 1962.

Anderson, Robert, and Ibrahim Fawzy, eds. *Egypt in 1800: Scenes from Napoleon's Description de l'Egypte.* London: Barrie & Jenkins, 1988.

Bonaparte's Proclamations As Recorded by Abd Al-Rahman Al-Jabarti. Cairo: Dar Al-Maaref, n.d.

Antes, John. *Observations on the Manners and Customs of the Egyptians, the Overflowing of the Nile and Its Effects; with Remarks on the Plague, and Other Subjects.* London: J. Stockdale, 1800.

Appel, Toby A. *The Cuvier-Geoffroy Debate: French Biology in the Decades Before Darwin.* New York: Oxford University Press, 1987.

Arberry, A. J. *Aspects of Islamic Civilization as Depicted in the Original Texts.* Ann Arbor: University of Michigan Press, 1967.

Ayalon, David. *Outsiders in the Lands of Islam: Mamluks, Mongols, and Eunuchs.* London: Variorum Reprints, 1988.

———. *Studies on the Mamluks of Egypt (1250–1517).* London: Variorum Reprints, 1977.

————. *Islam and the Abode of War: Military Slaves and Islamic Adversaries.* Aldershot, Great Britain; Brookfield, VT: Variorum, 1994.

Bainville, Jacques. *Napoleon.* Trans. Hamish Miles. London: J. Cape, 1932.

Baring, Sir Thomas. *A Bibliographical Account and Collation of La Description de l'Égypte: presented to the Library of the London Institution by Sir Thomas Baring, Baronet, president, with a list of the other donations made to that establishment from April 1837 to April 1838.* London: C. Skipper and East, 1838.

Barthélemy Saint-Hilaire, Jules. *Egypt and the Great Suez Canal. A Narrative of Travels.* London: R. Bentley, 1857.

Bell, Eric Temple. *Men of Mathematics.* New York: Simon and Schuster, 1937.

Bibliography of Scientific and Technical Literature Relating to Egypt, 1800–1900. (Comp. for the Survey department by C. Davies Sherborn). Cairo: Government Press, 1915.

Bierbrier, Morris L. *Historical Dictionary of Ancient Egypt.* Lanham, MD: Scarecrow Press, 1999.

————. *The Tomb-Builders of the Pharaohs.* London: British Museum Publications, 1982.

Bourrienne, Louis Antoine Fauvelet de. *Memoirs of Napoleon Bonaparte.* New York: F. A. Stokes Co. 1903.

Bourrienne, Louis Antoine Fauvelet de, and W. C. Armstrong, eds. *Bourrienne's Memoirs of Napoleon Bonaparte.* Hartford: Silas Andrus, 1853.

Boustany, Saladin. *The Journals of Bonaparte in Egypt, 1798–1801.* Cairo, Egypt: Al-Arab Bookshop, 1971.

Brewer, Douglas J., Donald B. Redford, and Susan Redford. *Domestic Plants and Animals: the Egyptian Origins.* Warminster, England: Aris & Phillips, 1994.

Bruce, James. *Travels to Discover the Source of the Nile, in the Years 1768, 1769, 1770, 1771, 1772 and 1773.* Edinburgh: J. Ruthven, 1790.

Byrd, Melanie. "The Napoleonic Institute of Egypt." Dissertation, University of Florida, Tallahassee, 1992.

Castelot, André. *Napoleon.* First ed. New York: Harper & Row, 1971.

Ceram, C. W. *Gods, Graves, and Scholars: The Story of Archaeology.* Second rev. ed. New York: Knopf, 1967.

Chandler, David G. *The Campaigns of Napoleon.* New York: Macmillan, 1966.

Copies of Original Letters from the Army of General Bonaparte in Egypt, Intercepted by the Fleet under the Command of Admiral Lord Nelson. With an English translation. Second ed. London: J. Wright, 1798.

Curl, James Stevens. *Egyptomania: The Egyptian Revival: A Recurring Theme in the History of Taste.* Manchester (England) and New York: Manchester University Press, 1994.

Dawson, Warren R., and Eric P. Uphill. *Who Was Who in Egyptology.* Third rev. ed. London: Egypt Exploration Society, 1995.

DeLaMater, Matt, Melanie Sue Byrd, and Yves Martin. "Napoleon in Egypt: A French Expedition Sails East to Strike at Britain." *Napoleon: International Journal of the French Revolution and Age of Napoleon* 13 (January 2001).

Denon, Vivant, and Edward Augustus Kendall. *Travels in Upper and Lower Egypt: During the Campaigns of General Bonaparte.* Second corrected ed. London: Cundee, 1803.

Description de l'Égypte. Ed. complete; trans. Chris Miller. Koln: Benedikt Taschen, 1994.

Dible, James Henry. *Napoleon's Surgeon.* London: Heinemann Medical, 1970.

Diodorus Siculus. *On Egypt.* Trans. Edwin Murphy. Jefferson, NC: McFarland, 1985.

Edwards, Holly, and Brian T. Allen. *Noble Dreams, Wicked Pleasures: Orientalism in America, 1870–1930.* Princeton: Princeton University Press in association with the Sterling and Francine Clark Art Institute, 2000.

Everett, Edward. "The Dendara Zodiac." *North American Review* 17 (1823): 221–41.

Fagan, Brian M. *The Rape of the Nile: Tomb Robbers, Tourists, and Archaeologists in Egypt.* Boulder, CO: Westview Press, 2004.

Farr, Florence. *Egyptian Magic.* Wellingborough, Great Britain: Aquarian, 1982.

Fournier, August, et al. *Napoleon the First, a Biography.* New York: Henry Holt & Co., 1903.

Fox, Robert. *Science, Industry, and the Social Order in Post-Revolutionary France.* Aldershot, Great Britain, and Brookfield, VT: Variorum, 1995.

Fregosi, Paul. *Dreams of Empire: Napoleon and the First World War, 1792–1815*. London: Hutchinson Ltd., 1989.

———. "Scientific Aspects of the French Egyptian Expedition, 1798–1801." *Proceedings of the American Philosophical Society* 20.1 (1995): 5–28.

Gillispie, Charles Coulston. *Science and Polity in France: The Revolutionary and Napoleonic Years*. Princeton: Princeton University Press, 2004.

Gillispie, Charles Coulston, ed. *Dictionary of Scientific Biography*. New York: Scribner, 1970–1980.

Gillispie, Charles Coulston, and Michel DeWachter, eds. *Monuments of Egypt: The Napeolonic Edition: the Complete Archaeological Plates from La Description de l'Égypte*. Princeton, NJ: Princeton Architectural Press in association with the Architectural League of New York, the J. Paul Getty Trust, 1987.

Godlewska, Anne. *Geography Unbound: French Geographic Science from Cassini to Humboldt*. Chicago: University of Chicago Press, 1999.

———. "Map, Text and Image: The Mentality of Enlightened Conquerors: a New Look at the Description de l'Égypte." *Transactions of the Institute of British Geographers, New Series* 20.1 (1995): 5–28.

Herivel, John. *Joseph Fourier: The Man and the Physicist*. Oxford: Clarendon Press, 1975.

Herold, J. Christopher. *Bonaparte in Egypt*. London: H. Hamilton, 1962.

Hillwood Art Museum. *Napoleon in Egypt*. Exhibition, August 17–September 30, 1990. Curator: Bob Brier. Brookville, NY: Hillwood Art Museum, 1990.

Houlihan, Patrick T. *Birds of Ancient Egypt*. Warminster, England: Aris & Philips, 1986.

Hourani, Albert Habib. *A History of the Arab Peoples*. Cambridge, MA: Belknap Press of Harvard University Press, 1991.

Humbert, Jean-Marcel, Michael Pantazzi, and Christiane Ziegler. *Egyptomania: Egypt in Western Art 1730–1930*. Exhibition: Paris, Musée du Louvre, 20 January–18 April 1994; Ottawa, National Gallery of Canada, 17 June–18 September 1994; Vienna, Kunsthistorisches Museum, 16 October 1994–29 January 1995. Ottawa: National Gallery of Canada. Paris: Réunion des Musées Nationaux, 1994.

Iverson, Eric. *The Myth of Egypt and its Hieroglyphics in European Tradition.* Princeton, NJ: Princeton University Press, 1993.

Jabarti, Abd al-Rahman, Louis Antoine Fauvelet de Bourrienne, and Edward W. Said. *Napoleon in Egypt: Al-Jabarti's Chronicle of the First Seven Months of the French Occupation, 1798.* Trans., Smuel Moreh. Princeton, NJ: Markus Wiener Publishers, 1993.

Jasanoff, Maya. *Edge of Empire: Lives, Culture and Conquest in the East, 1750–1850.* New York: Knopf, 2005.

Karabell, Zachary. *Parting the Desert: The Creation of the Suez Canal.* First ed. New York: Knopf, 2003.

Kinross, Baron Patrick Balfour. *Between Two Seas: The Creation of the Suez Canal.* New York: Morrow. 1969.

Kinross, Lord. *The Ottoman Centuries: The Rise and Fall of the Turkish Empire.* New York: Morrow. 1977.

Las Cases, Emmanuel-Auguste-Dieudonné, comte de. *The Life, Exile, and Conversations of the Emperor Napoleon. With Portraits and . . . Other Embellishments.* London: H. Colburn, 1835.

Lloyd, Christopher. *The Nile Campaign: Nelson and Napoleon in Egypt.* New York, Barnes and Noble, 1973.

Loring, John. "Egyptomania, the Nile Style." *The Connoisseur* 200.804 (1979 February): 114.

Lyons, Martyn. *France Under the Directory.* New York: Cambridge University Press, 1975.

Mayer, Leo Ary. *Mamluk Costume: A Survey.* Geneva: A. Kundig, 1952.

Moiret, Captain Joseph Marie, *Memoirs of Napoleon's Egyptian Expedition, 1798–1801.* Trans. and ed. by Rosemary Brindle. London: Greenhill Books, 2001.

Moorehead, Alan. *The Blue Nile.* First ed. New York: Harper & Row, 1962.

Murray, John. *A Handbook for Travellers in Lower and Upper Egypt.* Sixth ed., rev. on the spot. London: J. Murray, 1800.

Nowinski, Judith. *Baron Dominique Vivant Denon (1747–1825): Hedonist and Scholar in a Period of Transition.* Rutherford, NJ: Fairleigh Dickinson University Press, 1970.

Otter, William. *The Life and Remains of Edward Daniel Clarke, Professor of Mineralogy in the University of Cambridge.* London: G. Cowie and Co., 1825.

Parkinson, Richard B., et al. *Cracking Codes: The Rosetta Stone and Decipherment.* London: British Museum Press, 1999.

Paton, A. A. *A History of the Egyptian Revolution, from the Period of the Mamelukes to the Death of Mohammed Ali.* Second ed., enl. London: Trubner & Co., 1870.

Picturing the Middle East: A Hundred Years of European Orientalism. New York: Dahesh Museum, 1996.

Porterfield, Todd. "Egyptomania!" *Art in America* 82.11 (1994 November): 84.

Pratt, Mary Louise. *Imperial Eyes: Travel Writing and Transculturation.* London and New York: Routledge 1992.

Raymond, André. *Cairo.* Trans. Willard Wood. Cambridge, MA: Harvard University Press, 2000.

Reid, Donald Malcolm. *Whose Pharaohs?: Archaeology, Museums, and Egyptian National Identity from Napoleon to World War I.* Berkeley: University of California Press, 2002.

Richardson, Robert G. *Larrey: Surgeon to Napoleon's Imperial Guard.* London: Murray, 1974.

Rodenbeck, Max. *Cairo: The City Victorious.* First Amer. ed. New York: Knopf, 1999.

Said, Edward W. *Reflections on Exile and Other Essays.* Cambridge, MA: Harvard University Press, 2000.

Sayyid-Marsot, Afaf Lutfi. *Women and Men in Late Eighteenth Century Egypt.* First ed. Austin: University of Texas Press, 1995.

Silberman, Neil Asher. "That Miserable Fort!" *Military History Quarterly* 3.2 (1991): 62–72.

Siliotti, Alberto. *The Discovery of Ancient Egypt.* Cairo: American University in Cairo Press, 1998.

Sonnini, C. S. *Travels in Upper and Lower Egypt: Undertaken by Order of the Old Government of France.* Trans. Henry Hunter. London: J. Stockdale, 1807.

Trafton, Scott. *Egypt Land: Race and Nineteenth-Century American Egyptomania.* Durham: Duke University Press, 2004.

Trigger, Bruce G. *A History of Archeological Thought.* Cambridge and New York: Cambridge University Press, 1989.

Truman, Charles. *Sèvres Egyptian Service, 1810–12*. London: Victoria and Albert Museum, 1982.

Tulard, Jean. *Napoleon: The Myth of the Saviour.* Trans. Teresa Waugh. London: Weidenfeld and Nicolson, 1984.

Volney, C.-F. *Travels Through Syria and Egypt, in the Years 1783, 1784 and 1785, Containing the Present Natural and Political State of Those Countries, Their Productions, Arts, Manufacturers and Commerce.* London: G.G.J. and J. Robinson, 1787.

Wittman, William. *Travels in Turkey, Asia-Minor, Syria, and Across the Desert into Egypt: During the Years 1799, 1800, and 1801, in Company with the Turkish Army, and the British Military Mission: to Which Are Annexed, Observations on the Plague, and on the Diseases Prevalent in Turkey, and a Meteorological Journal.* London: Richard Phillips, 1803.

Wortham, John David. *The Genesis of British Egyptology, 1549–1906.* First ed. Norman: University of Oklahoma Press, 1971.

SOURCES IN FRENCH

Arago, François. *Oeuvres complètes de François Arago, secrétaire perpétuel de l'academie des sciences. Notices biographiques. Malus.* Tome 3, Volume 3. Paris: Gide and J. Baudry editeurs. 1855.

Description de l'Égypte, ou Recueil des observations et des recherches qui ont été faites en Egypte pendant l'Expédition de l'Armée Française. 2nd (Panckoucke) edition. All Volumes. Paris, Imprimerie de C.L.F. Panckoucke, 1821–1829.

Bernoyer, François, Chef de l'atelier d'habillement de l'Armée d'Orient. *Avec Bonaparte en Égypte et en Syrie: 1798–1800.* 19 lettres inédites retrouvées et présentées par Christian Tortel. Editions Curandera, 1981.

Bret, Patrice, l'astronome Nicolas-Antoine Nouet (1740–1811), membre de l'Institut d'Égypte, directeur de la carte de Savoie. *Les Scientifiques et la Montagne, Actes du 116 Congrès National des sociétés savantes (Chambéry, 29 avril–4 mai 1991), Section d'histoire des sciences et des techniques.* Paris: Editions du CTHS, 1993. 119–47.

Bret, Patrice, ed. *L'Expedition d'Égypte, une entreprise lumières, 1798–1801.* Actes de Colloques, 1999.

Cahn, Théophile. *LaVie et l'oeure d' Étienne Geoffroy saint-Hilaire.* Paris: Presses universitaires de France, 1962.

Daressy, Georges. "L'Ingénieur Girard et l'Institut d'Egypte" *Bulletin de l'Institut égyptien,* 5e série, tome XII, 1918, 13–32.

Devilliers du Terrage, E. *Journal et Souvenirs sur l'Expédition d'Egypte (1798–1801). Mis en ordre et publiés par le Baron Marc Devilliers du Terrage.* Paris: Plon, 1899.

Desgenettes, René. *Histoire médicale de l'armée d'Orient.* 1802.

Dhombres, Jean. *L'Esprit de Géométrie en Egypte. Monge et Fourier et Jomard: de la Science conquérante à la science positivée.*

Gazel, Louis. *Le baron DesGenettes (1762–1837): Notes biographiques.* [Thesis for a doctorate in medicine.] Paris: Henry Paulin & Cie, Editeurs, 1912.

Geoffroy Saint-Hilaire, Étienne. *Lettres écrites d'Egypte à Cuvier, Jussieu, Lacépède, Monge, Desgenettes, Redouté jeune, Norry, Etc., aux professeurs du Muséum et à sa famille.* Recueillies et publiées avec une préface et des notes par le Docteur E.-T. Hamy. Paris: Librairie Hachette et Cie, 1901.

Geoffroy Saint-Hilaire, Isidore. *Vie, travaux et doctrine scientifique d'Étienne Geoffroy Saint-Hilaire.* Paris: P. Bertrand, Éditeur, 1847. Electronic version (BNF), 1995.

Goby, Jean-Edouard. "Antoine-François-Ernest Coquebert de Monbret, bibliothecaire . . ." *Bulletin de l'institut d'Egypte,* tome XXXI, 1948–1949, 77–87.

———. *Un compagnon de Bonaparte en Egypte: Dubois-Aymé. Cahiers d'histoire égyptienne.* Série III, fasc. 3, mars 1951, 221–54.

Jollois, Prosper. *Journal d'un ingénieur: Attaché à l'expédition d'Égypte: 1798–1802.* [Notes de voyage et d'archéologie rédigées par Prosper Jollois, Ingénieur des Ponts et Chaussées, Membre de la Commission des Sciences et Arts, avec des fragments tirés des journaux de Fourier, Jomard, Delile, Saint-Génis, Descostils, Balzac et Coraboeuf.] Publié par P. Lefèvre-Pontalis. Ernest Leroux, éditeur, Paris, 1904. [Tome 6. Bibliothèque égyptologique contenant les œuvres des égyptologues français dispersées dans divers recueils et qui n'ont pas encore été réunies jusqu'à ce jour. Publiée sous la direction de G. Maspero, Membre de l'Institut, directeur d'études à l'Ecole pratique des Hautes-Études, professeur au Collège de France.]

Jomard, Edme-François. *Conté.* Paris: E. Thunot et Cie, 1849.

Lacroix, A., and Daressy, G. *Dolomieu en Egypte (30 juin 1798–10 mars 1799): Mémoires présentés à l'Institut d'Égypte et publiés sous les auspices de sa Majesté Fouad Ier, Roi d'Egypte.* Volume 3. Le Caire: Imprimerie de l'Institut français d'archéologie orientale, 1922.

Laissus, Yues. *L Égypte, une aventure savante avec Bonaparte, Kléber, Menou 1798–1801.* Paris: Librairie Arithème Fayard, 1998.

LeGrain, Georges. "Guillame-André Villoteau, musicographe de l'Expédition française d'Egypte (1759–1839)." *Bulletin de l'Institut égyptien,* cinquième série, Tome XI, 1917, 1–30.

Malus, Étienne. *L'Agenda de Malus: Souvenirs de l'expeditions d'egypte. 1798–1801.* Publié et annoté par Général Thoumas. Paris: Honoré Champion Librairie, 1892.

Pallary, Paul. "Marie Jules-César Savigny: Sa vie et son œuvre. Première partie.—La vie de Savigny." Mémoires présentés à l'Institut d'Égypte et publiés sous les auspices de sa Majesté Fouad Ier, Roi d'Egypte. Vol 17. Le Caire: Imprimerie de l'Institut français d'archéologie orientale, 1931.

Reybaud, Louis, et al. *Histoire scientifique et militaire de l'expédition française en Egypte,* Tome 3. Paris: A.-J. Dénain, libraire-éditeur, 1830–1836.

Sadoun-Goupil, Michelle. *Le Chimiste Claude-Louis Berthollet (1748–1822): Sa Vie and Son Oeuvre.* Paris: Librairie Philosophique J. Vrin, 1977.

La Sabretache. "La mission d'Ernest Coquebert de Montbret" *Carnet de La Sabretache: Revue d'histoire militaire.* Publiée par la société La Sabretache. Number 414, juin 1956, 43–86.

INDEX

Index

Index